# Environmental Health Criteria 134

# CADMIUM

First draft prepared by Dr L. Friberg and Dr C.G. Elinder
(Karolinska Institute, Sweden) and Dr. T. Kjellström
(University of Auckland, New Zealand)

Published under the joint sponsorship of
the United Nations Environment Programme,
the International Labour Organisation,
and the World Health Organization



World Health Organization
Geneva, 1992

The **International Programme on Chemical Safety (IPCS)** is a joint venture of the United Nations Environment Programme, the International Labour Organisation, and the World Health Organization. The main objective of the IPCS is to carry out and disseminate evaluations of the effects of chemicals on human health and the quality of the environment. Supporting activities include the development of epidemiological, experimental laboratory, and risk-assessment methods that could produce internationally comparable results, and the development of manpower in the field of toxicology. Other activities carried out by the IPCS include the development of know-how for coping with chemical accidents, coordination of laboratory testing and epidemiological studies, and promotion of research on the mechanisms of the biological action of chemicals.

WHO Library Cataloguing in Publication Data

Cadmium

(Environmental health criteria ; 134)

1.Cadmium - adverse effects 2.Cadmium - toxicity
3.Environmental exposure 4.Environmental pollutants
I.Series

ISBN 92 4 157134 9 (NLM Classification: QV 290)
ISSN 0250-863X

The World Health Organization welcomes requests for permission to reproduce or translate its publications, in part or in full. Applications and enquiries should be addressed to the Office of Publications, World Health Organization, Geneva, Switzerland, which will be glad to provide the latest information on any changes made to the text, plans for new editions, and reprints and translations already available.



The designations employed and the presentation of the material in this publication do not imply the expression of any opinion whatsoever on the part of the Secretariat of the World Health Organization concerning the legal status of any country, territory, city or area or of its authorities, or concerning the delimitation of its frontiers or boundaries.

The mention of specific companies or of certain manufacturers' products does not imply that they are endorsed or recommended by the World Health Organization in preference to others of a similar nature that are not mentioned. Errors and omissions excepted, the names of proprietary products are distinguished by initial capital letters.

Printed in Finland

92/9233 — Vammala — 5500

# CONTENTS

## WHO TASK GROUP ON ENVIRONMENTAL HEALTH CRITERIA FOR CADMIUM

*Members*

Professor K.A. Bustueva, Communal Hygiene, Central Institute for Advanced Medical Training, Moscow, USSR

Dr S.V. Chandra, Industrial Toxicology Research Centre, Mahatma Gandhi Marg, Lucknow, India

Dr M.G. Cherian, Department of Pathology, University of Western Ontario, London, Ontario, Canada *(Joint Rapporteur)*

Dr B.A. Fowler, School of Medicine, University of Maryland, Baltimore, Maryland, USA *(Joint Rapporteur)*

Dr R.A. Goyer, Department of Pathology, University of Western Ontario, London, Ontario, Canada *(Chairman)*

Professor G. Kazantzis, London School of Hygiene and Tropical Medicine, University of London, London, United Kingdom

Professor G. Nordberg, Department of Environmental Medicine, University of Umea, Umea, Sweden

Dr J. Parizek, Czechoslovak Academy of Sciences, Institute of Physiology, Videnska, Prague, Czechoslovakia

Dr I. Shigematsu, Radiation Effects Research Foundation, Hijiyama Park, Minami-Ku, Hiroshima, Japan *(Vice-Chairman)*

Dr M.J. Thun, Division of Epidemiology and Statistics, American Cancer Society, Atlanta, Georgia, USA

*Observers*

Professor K. Nogawa, Department of Hygiene, Chiba University School of Medicine, Chiba, Japan

Dr K. Nomiyama, Department of Environmental Health, Jichi Medical School, Minamikawachi-Machi Kawachi-Gun, Tochigi-Ken, Japan

*Secretariat*

Dr G.C. Becking, International Programme on Chemical Safety, Interregional Research Unit, World Health Organization, Research Triangle Park, North Carolina, USA *(Secretary)*

Dr L. Friberg, Karolinska Institute, Department of Environmental Hygiene, Stockholm, Sweden

Dr C.G. Elinder, Section for Renal Medicine, Department of Internal Medicine, Karolinska Hospital, Stockholm, Sweden

## NOTE TO READERS OF THE CRITERIA MONOGRAPHS

Every effort has been made to present information in the criteria monographs as accurately as possible without unduly delaying their publication. In the interest of all users of the environmental health criteria monographs, readers are kindly requested to communicate any errors that may have occurred to the Manager of the International Programme on Chemical Safety, World Health Organization, Geneva, Switzerland, in order that they may be included in corrigenda.

* * *

A detailed data profile and a legal file can be obtained from the International Register of Potentially Toxic Chemicals, Palais des Nations, 1211 Geneva 10, Switzerland (Telephone No. 7988400 or 7985850).

## ENVIRONMENTAL HEALTH CRITERIA FOR CADMIUM

A WHO Task Group on Environmental Health Criteria for Cadmium met in Geneva from 27 November to 1 December 1989. Dr M. Mercier, Manager, IPCS, opened the meeting on behalf of the heads of the three IPCS cooperating organizations (UNEP/ILO/WHO). The Task Group reviewed and revised the draft criteria document and made an evaluation of the risks to human health from exposure to cadmium.

The first draft of this monograph, which was reviewed by a Working Group in January 1984, was prepared by Dr L. Friberg and Dr C.G. Elinder (Karolinska Institute, Stockholm, Sweden), and Dr T. Kjellström (University of Auckland, New Zealand)[1]. Based on the discussions of the Working Group, recent scientific data, and comments from the IPCS Contact Points, a Task Group draft was prepared by Dr R. Goyer (University of Western Ontario, Canada).

The Secretariat wishes to acknowledge the contributions made by Professor K. Tsuchiya (Keio University, Tokyo, Japan), Dr M. Piscator (Karolinska Institute), Dr G.F. Nordberg (University of Umea, Sweden), and Professor R. Lauwerys (University of Louvain, Brussels, Belgium) for their preparation and review of earlier draft document on cadmium, which assisted greatly in the preparation of this monograph.

Dr G.C. Becking (Interregional Research Unit) and Dr P.G. Jenkins (IPCS Central Unit) were responsible for the overall scientific content and technical editing, respectively, of this monograph. The efforts of all who helped in the preparation and finalization of the document are gratefully acknowledged.

[1] Present affiliation: Division of Environmental Health, World Health Organization

## ABBREVIATIONS

| | |
|---|---|
| AAS | atomic absorption spectrometry |
| CC | critical concentration |
| CI | confidence interval |
| EEC | European Economic Community |
| ETA | electrothermal atomization |
| GESAMP | Group of Experts on the Scientific Aspects of Marine Pollution |
| GFR | glomerular filtration rate |
| GOT | glutamic-oxaloacetic transaminase |
| GPT | glutamic-pyruvic transaminase |
| ICD | International Classification of Diseases |
| IDMS | isotope dilution mass spectrometry |
| IU | international units |
| LDH | lactate dehydrogenase |
| LMW | low molecular weight |
| MMAD | mass median aerodynamic diameter |
| PCV | packed-cell volume |
| PMR | proportional mortality rate |
| PMSG | pregnant mare serum gonadotrophin |
| RBP | retinal binding protein |
| RIA | radio-immuno assay |
| SMR | standard mortality ratio |
| TRP | tubular reabsorption of phosphate |
| XRF | X-ray-generated atomic fluorescence |

# PREFACE

The definitions of terms used in this monograph were derived from the meeting of the Scientific Committee on the Toxicology of Metals, Permanent Commission and International Association on Occupational Health, in Tokyo in 1974 (Task Group on Metal Toxicity, 1976). The term "critical concentration" in an organ was defined as "the concentration of a metal in an organ at the time any of its cells reaches a concentration at which adverse functional changes, reversible or irreversible, occur in the cell". These first adverse changes would be the "critical effect". The critical concentration is thus established on an individual level and varies between individuals. The term "critical organ" was defined as "that particular organ which first attains the critical concentration of a metal under specified circumstances of exposure and for a given population".

The dose-response relationship expressing the occurrence rate (response) of the particular effect as a function of metal concentration in the critical organ, displays the frequency distribution of individual critical concentrations. In risk estimations it is thus essential to define the variability of the critical concentration among a population or specific group of people.

The term that was chosen to predict the variability of the critical concentration of cadmium occurring in a particular group of people is the predicted prevalence of the critical concentration. For example, the critical concentration 5 ($CC_5$) would be the concentration at which 5% of the population had reached their individual critical concentrations, and the $CC_{50}$ would be the critical concentration occurring in 50% of a defined group of people. The term "critical concentration" is synonymous with the term "population critical concentration" used in the WHO publication on Evaluation of Certain Food Additives and Contaminants (1989).

The critical concentrations and the dose-response relationships are very much dependent on the definition of critical effect. The early effects of cadmium on the kidney can be measured as an increased urinary excretion of low molecular weight (LMW) proteins. An operational definition is needed to create a cut-off point above which the proteinuria indicates an "adverse functional change". Different studies of cadmium effects have used different operational definitions, which has made it difficult to merge the data into a dose-response relationship. Examples of these problems are given in section 8.3.2. The relationship between dose and different types of effect or different severities of the same effect is called the dose-effect relationship.

In animal studies, the individual critical concentrations have not been calculated. Both dose and effect data are based on groups of animals, and these groups are usually rather small (section 7). Few animal studies attempt to quantitatively measure the dose-response relationships within the group (section 7.2.1.4). The reports of effects occurring at a certain concentration of cadmium in the kidney cortex may therefore best be interpreted as the concentration at which 50% or more of the animals suffered the effect. A 5-10% response will occur at lower cadmium concentrations.

The effects of cadmium on the environment are discussed in Environmental Health Criteria 135: Cadmium - Environmental Aspects (WHO, in press).

# 1. SUMMARY AND CONCLUSIONS

## 1.1 Identity, physical and chemical properties, and analytical methods

Several methods are available for the determination of cadmium in biological materials. Atomic absorption spectrometry is the most widely used, but careful treatment of samples and correction for interference is needed for the analysis of samples with low cadmium concentrations. It is strongly recommended that analysis be accompanied by a quality assurance programme. At present, it is possible under ideal circumstances to determine concentrations of about 0.1 μg/litre in urine and blood and 1-10 μg/kg in food and tissue samples.

## 1.2 Sources of human and environmental exposure

Cadmium is a relatively rare element and current analytical procedures indicate much lower concentrations of the metal in environmental media than did previous measurements. At present, it is not possible to determine whether human activities have caused a historic increase in cadmium levels in the polar ice caps.

Commercial cadmium production started at the beginning of this century. The pattern of cadmium consumption has changed in recent years with significant decreases in electroplating and increases in batteries and specialized electronic uses. Most of the major uses of cadmium employ cadmium in the form of compounds that are present at low concentration; these features constrain the recycling of cadmium. Restrictions on certain uses of cadmium imposed by a few countries may have widespread impact on these applications.

Cadmium is released to the air, land, and water by human activities. In general, the two major sources of contamination are the production and consumption of cadmium and other non-ferrous metals and the disposal of wastes containing cadmium. Areas in the vicinity of non-ferrous mines and smelters often show pronounced cadmium contamination.

Increases in soil cadmium content result in an increase in the uptake of cadmium by plants; the pathway of human exposure from agricultural crops is thus susceptible to increases in soil cadmium. The uptake by plants from soil is greater at low soil pH.

Processes that acidify soil (e.g., acid rain) may therefore increase the average cadmium concentrations in foodstuffs. The application of phosphate fertilizers and atmospheric deposition are significant sources of cadmium input to arable soils in some parts of the world; sewage sludge can also be an important source at the local level. These sources may, in the future, cause enhanced soil and hence crop cadmium levels, which in turn may lead to increases in dietary cadmium exposure. In certain areas, there is evidence of increasing cadmium content in food.

Edible free-living food organisms such as shellfish, crustaceans, and fungi are natural accumulators of cadmium. As in the case of humans, there are increased levels of cadmium in the liver and kidney of horses and some feral terrestrial animals. Regular consumption of these items can result in increased exposure. Certain marine vertebrates contain markedly elevated renal cadmium concentrations, which, although considered to be of natural origin, have been linked to signs of kidney damage in the organisms concerned.

## 1.3 Environmental levels and human exposure

The major route of exposure to cadmium for the non-smoking general population is via food; the contribution from other pathways to total uptake is small. Tobacco is an important source of cadmium uptake in smokers. In contaminated areas, cadmium exposure via food may be up to several hundred μg/day. In exposed workers, lung absorption of cadmium following inhalation of workplace air is the major route of exposure. Increased uptake can also occur as a consequence of contamination of food and tobacco.

## 1.4 Kinetics and metabolism in laboratory animals and humans

Data from experimental animals and humans have shown that pulmonary absorption is higher than gastrointestinal absorption. Depending on chemical speciation, particle size, and solubility in biological fluids, up to 50% of the inhaled cadmium compound may be absorbed. The gastrointestinal absorption of cadmium is influenced by the type of diet and nutritional status. The nutritional iron status appears to be of particular importance. On average, 5% of the total oral intake of cadmium is absorbed, but individual values range from less than 1% to more than 20%. There is a maternal-fetal gradient of cadmium. Although cadmium accumulates in the placenta, transfer to the fetus is low.

Cadmium absorbed from the lungs or the gastrointestinal tract is mainly stored in the liver and kidneys, where more than half of the body burden will be deposited. With increasing exposure intensity, an increasing proportion of the absorbed cadmium is stored in the liver. Excretion is normally slow, and the biological half-time is very long (decades) in the muscles, kidneys, liver, and whole body of humans. The cadmium concentrations in most tissues increase with age. Highest concentrations are generally found in the renal cortex, but excessive exposures may lead to higher concentrations in the liver. In exposed people with renal damage, urinary excretion of cadmium increases and so the whole body half-time is shortened. The renal damage leads to losses of cadmium from the kidney, and the renal concentrations of cadmium will eventually be lower than in people with similar exposure but without renal damage.

Metallothionein is an important transport and storage protein for cadmium and other metals. Cadmium can induce metallothionein synthesis in many organs including the liver and kidney. The binding of intracellular cadmium to metallothionein in tissues protects against the toxicity of cadmium. Cadmium not bound to metallothionein may therefore play a role in the pathogenesis of cadmium-related tissue injury. The speciation of other cadmium complexes in tissues or biological fluids is unknown.

Urinary excretion of cadmium is related to body burden, recent exposure, and renal damage. In people with low exposure, the urine cadmium level is mainly related to the body burden. When cadmium-induced renal damage has occurred, or even without renal damage if exposure is excessive, urinary excretion increases. Cadmium-exposed people with proteinuria generally have higher cadmium excretion than such people without proteinuria. After high exposure ceases, the urine cadmium level will decrease even though renal damage persists. The interpretation of urinary cadmium is thus dependent on a number of factors. Gastrointestinal excretion is approximately equal to urinary excretion but cannot be easily measured. Other excretory routes such as lactation, sweating or placental transfer are insignificant.

The level of cadmium in faeces is a good indicator of recent daily intake from food in the absence of inhalation exposure. Cadmium in blood occurs mainly in the red blood cells, and the plasma concentrations are very low. There are at least two compartments in blood, one related to recent exposure with a half-time of about 2-3 months, and one which is probably related to body burden with a half-time of several years.

## 1.5 Effects on laboratory mammals

High inhalation exposures cause lethal pulmonary oedema. Single high-dose injection gives rise to testicular and non-ovulating ovarian necrosis, liver damage, and small vessel injury. Large oral doses damage the gastric and intestinal mucosa.

Long-term inhalation exposure and intratracheal administration give rise to chronic inflammatory changes in the lungs, fibrosis, and appearances suggestive of emphysema. Long-term parenteral or oral administration produces effects primarily on the kidneys, but also on the liver and the haematopoietic, immune, skeletal, and cardiovascular systems. Skeletal effects and hypertension have been induced in certain species under defined conditions. The occurrence of teratogenic effects and placental damage depends on the stage of gestation at which exposure occurs, and may involve interactive effects with zinc.

Of greatest relevance to human exposure are the acute inhalation effects on the lung and the chronic effects on the kidney. Following long-term exposure, the kidney is the critical organ. The effects on the kidney are characterized by tubular dysfunction and tubular cell damage, although glomerular dysfunction may also occur. A consequence of renal tubular dysfunction is a disturbance of calcium and vitamin D metabolism. According to some studies, this has led to osteomalacia and/or osteoporosis, but these effects have not been confirmed by other studies. A direct effect of cadmium on bone mineralization cannot be excluded. The toxic effects of cadmium in experimental animals are influenced by genetic and nutritional factors, interactions with other metals, particularly zinc, and pretreatment with cadmium, which may be related to the induction of metallothionein.

In 1976 and 1987, the International Agency for Research on Cancer accepted as sufficient the evidence that cadmium chloride, sulfate, sulfide, and oxide can give rise to injection site sarcomas in the rat and, for the first two compounds, induce interstitial cell tumours of the testis in rats and mice, but found oral studies inadequate for evaluation. Long-term inhalation studies in rats exposed to aerosols of cadmium sulfate, cadmium oxide fumes and cadmium sulfate dust demonstrated a high incidence of primary lung cancer with evidence of a dose-response relationship. However, this has not so far been demonstrated in other species. Studies on the genotoxic effects of cadmium have given discordant results.

## 1.6 Effects on humans

High inhalation exposure to cadmium oxide fume results in acute pneumonitis with pulmonary oedema, which may be lethal. High ingestion exposure of soluble cadmium salts causes acute gastroenteritis.

Long-term occupational exposure to cadmium has caused severe chronic effects, predominantly in the lungs and kidneys. Chronic renal effects have also been seen among the general population.

Following high occupational exposure, lung changes are primarily characterized by chronic obstructive airway disease. Early minor changes in ventilatory function tests may progress, with continued cadmium exposure, to respiratory insufficiency. An increased mortality rate from obstructive lung disease has been seen in workers with high exposure, as has occurred in the past.

The accumulation of cadmium in the renal cortex leads to renal tubular dysfunction with impaired reabsorption of, for instance, proteins, glucose, and amino acids. A characteristic sign of tubular dysfunction is an increased excretion of low molecular weight proteins in urine. In some cases, the glomerular filtration rate decreases. Increase in urine cadmium correlates with low molecular weight proteinuria and in the absence of acute exposure to cadmium may serve as an indicator of renal effect. In more severe cases there is a combination of tubular and glomerular effects, with an increase in blood creatinine in some cases. For most workers and people in the general environment, cadmium-induced proteinuria is irreversible.

Among other effects are disturbances in calcium metabolism, hypercalciuria, and formation of renal stones. High exposure to cadmium, most probably in combination with other factors such as nutritional deficiencies, may lead to the development of osteoporosis and/or osteomalacia.

There is evidence that long-term occupational exposure to cadmium may contribute to the development of cancer of the lung but observations from exposed workers have been difficult to interpret because of confounding factors. For prostatic cancer, evidence to date is inconclusive but does not support the suggestion from earlier studies of a causal relationship.

At present, there is no convincing evidence for cadmium being an etiological agent of essential hypertension. Most data speak against a blood pressure increase due to cadmium and there is no evidence of an increased mortality due to cardiovascular or cerebrovascular disease.

Data from studies on groups of occupationally exposed workers and on groups exposed in the general environment show that there is a relationship between exposure levels, exposure durations, and the prevalence of renal effects.

An increased prevalence of low molecular weight proteinuria in cadmium workers after 10-20 years of exposure to cadmium levels of about 20-50 $\mu g/m^3$ has been reported.

In polluted areas of the general environment, where the estimated cadmium intake has been 140-260 $\mu$g/day, effects in the form of increased low molecular weight proteinuria have been seen in some individuals following long-term exposure. More precise dose-response estimates are given in section 8.

## 1.7 Evaluation of human health risks

### 1.7.1 *Conclusions*

The kidney is considered the critical target organ for the general population as well as for occupationally exposed populations. Chronic obstructive airway disease is associated with long-term high-level occupational exposure by inhalation. There is some evidence that such exposure to cadmium may contribute to the development of cancer of the lung but observations from exposed workers have been difficult to interpret because of confounding factors.

#### 1.7.1.1 *General population*

Food-borne cadmium is the major source of exposure for most people. Average daily intakes from food in most areas not polluted with cadmium are between 10-40 $\mu$g. In polluted areas it has been found to be several hundred $\mu$g per day. In non-polluted areas, uptake from heavy smoking may equal cadmium intake from food.

Based on a biological model, an association between cadmium exposure and increased urinary excretion of low molecular weight proteins has been estimated to occur in humans with a life-long daily intake of about 140-260 μg cadmium, or a cumulative intake of about 2000 mg or more.

#### 1.7.1.2 *Occupationally exposed population*

Occupational exposure to cadmium is mainly by inhalation but includes additional intakes through food and tobacco. The total cadmium level in air varies according to industrial hygiene practices and type of workplace. There is an exposure-response relationship between airborne cadmium levels and proteinuria. An increase in the prevalence of low molecular weight proteinuria may occur in workers after 10-20 years of exposure to cadmium levels of about 20-50 $\mu g/m^3$. *In vivo* measurement of cadmium in the liver and kidneys of people with different levels of cadmium exposure have shown that about 10% of workers with a kidney cortex level of 200 mg/kg and about 50% of people with a kidney cortex level of 300 mg/kg would have renal tubular proteinuria.

# 2. IDENTITY, PHYSICAL AND CHEMICAL PROPERTIES, AND ANALYTICAL METHODS

This monograph covers cadmium and its inorganic compounds alone, since there is no evidence that organocadmium compounds (where the metal is bound covalently to carbon) occur in nature. Although cadmium may bind to proteins and other organic molecules and form salts with organic acids (e.g., cadmium stearate), in these forms it is regarded as inorganic.

The mobility of cadmium in the environment and the effects on the ecosystem depend to a large extent on the nature of its compounds.

Since this monograph evaluates only the health hazards for humans (and not those for the environment), only chemical data on cadmium compounds relevant for such an evaluation are included. Data on cadmium compounds occurring in or toxic to lower animals and plants are reviewed in Environmental Health Criteria 135: Cadmium - Environmental Aspects (WHO, in press).

## 2.1 Physical and chemical properties

Cadmium (atomic number 48; relative atomic mass 112.40) is a metal that belongs, together with zinc and mercury, to group IIb in the Periodic Table. Naturally-occurring isotopes are 106 (1.22%), 108 (0.88%), 110 (12.39%), 111 (12.75%), 112 (24.07%), 113 (12.26%), 114 (28.86%), and 116 (7.50%) (Weast, 1974).

Cadmium has a relatively high vapour pressure. Its vapour is oxidized rapidly in air to produce cadmium oxide. When reactive gases or vapour, such as carbon dioxide, water vapour, sulfur dioxide, sulfur trioxide or hydrogen chloride are present, cadmium vapour reacts to produce cadmium carbonate, hydroxide, sulfite, sulfate or chloride, respectively. These compounds may be formed in stacks and emitted to the environment. An example of these reactions during cadmium emissions from coal-fired power plants is described by Kirsch et al. (1982).

Some cadmium compounds, such as cadmium sulfide, carbonate, and oxide, are practically insoluble in water. There is, however, a lack of data on the solubility of these compounds in biological fluids, e.g., in the gastrointestinal tract and lung. These water-insoluble compounds can be changed to water-soluble salts in nature under the influence of oxygen and acids; cadmium sulfate, nitrate, and halides are water-soluble. Most of the cadmium found in mammals, birds, and fish is probably bound to protein molecules.

The speciation of cadmium in soil, plants, animal tissues, and foodstuffs may be of importance for the evaluation of the health hazards associated with areas of cadmium contamination or high cadmium intake. For example, although soil cadmium levels in Shipham, United Kingdom, were found to be very much higher than in Toyama, Japan, cadmium uptake by edible plants in Shipham was only a small fraction of that in Toyama (Tsuchiya, 1978; Sherlock et al., 1983). Very few data on the occurrence and speciation of cadmium compounds in nature are available.

## 2.2 Analytical methods

Only a few nanograms (or even less) of cadmium may be present in collected samples of air and water, whereas hundreds of micrograms may be present in small samples of kidney, sewage sludge, and plastics. Different techniques are therefore required for the collection, preparation, and analysis of the samples.

In general, the technique available for measuring cadmium in the environment and in biological materials cannot differentiate between the different compounds. With special separation techniques, cadmium-containing proteins can be isolated and identified. In most studies to date, the concentration or amount of cadmium in water, air, soil, plants, and other environmental or biological materials has been determined as the element.

### 2.2.1 Collection and preparation of samples

The degree of uncertainty in any health risk assessment of cadmium based on the analysis of environmental or biological samples depends on how representative the samples are. Each type of material has specific problems in this respect, and each study should include an evaluation of the sampling procedures utilized. For example, the measurement of cadmium in workplace air can be made with "static" samples or "personal" samples. The latter supposedly gives a better estimate of true exposure levels. When both are measured, personal samples usually give higher results, indicating that static samples may underestimate the exposure.

For the collection of samples, standard trace element methods can generally be used (LaFleur, 1976; Behne, 1980). The amount of material needed for analysis varies according to the sensitivity of the analytical methods and the cadmium concentration in the material. During recent years, methods have improved and usually smaller amounts (ml or g) of biological materials are now needed than those required previously.

In the handling and storage of samples, particularly liquid samples, special care must be taken to avoid contamination. Coloured materials in containers, especially plastics and rubber, should be avoided. Contamination of blood samples has been reported when blood was collected in certain types of evacuated blood collection tubes (Nackowski et al., 1977; Nise & Vesterberg, 1978). Disposable coloured micropipette tips have been found to contaminate acid solutions with cadmium (Salmela & Vuori, 1979).

Glass and transparent cadmium-free polyethylene, polypropylene or teflon containers are usually considered as suitable for storing samples. All containers and glassware should be precleaned in dilute nitric acid and deionized water. Water samples or standards with low cadmium concentrations should be stored for only a short period of time in order to avoid possible adsorption of cadmium on the container wall. However, experiments carried out within the UNEP/WHO programme (Vahter, 1982; Friberg & Vahter, 1983) using haemolysed blood samples spiked with $^{109}$Cd showed that, if properly handled, blood can be stored at room temperature for several months without any change in the cadmium concentration. Some solutions, such as urine, should be acidified to prevent precipitation of salts, thus ensuring that the cadmium remains in solution.

To prepare samples for analysis, inorganic solid samples (such as soil or dust samples) are usually dissolved in nitric acid or other acids. Organic samples need to be subjected to wet ashing (digestion) or dry ashing. Wet ashing, i.e. heating under reflux with nitric acid followed by the addition of sulfuric or perchloric acid, is an adequate method for the digestion of most organic and biological samples. Heating with perchloric acid is usually avoided in modern methods because of the explosive nature of the fumes. Biological samples may also be dissolved using tetramethylammonium hydroxide (Kaplan et al., 1973).

Dry ashing can also be used without significant losses of cadmium, provided that the temperature is kept at or below 450 °C (Kjellström et al., 1974; Koirtyohann & Hopkins, 1976). Low-temperature (about 100 °C) dry ashing at a high oxygen concentration has also been used successfully (Gleit, 1965).

### 2.2.2 Separation and concentration

Some biological samples such as kidneys contain relatively high concentrations of cadmium; this makes it possible to analyse without significant interference from other compounds. Dry ashing, followed by dissolving the ash in acid, is sometimes sufficient for analysis by atomic absorption spectrometry and other modern methods. When the cadmium concentration is low, special treatment is sometimes needed. The procedures for separating cadmium from interfering compounds and concentrating the samples are very important steps in obtaining adequate results.

One technique for the solvent extraction of cadmium, which has been widely used, is based on the APDC/MIBK system, where ammonium pyrrolidine dithiocarbamate chelate (APDC) is extracted into methyl isobutyl ketone (MIBK) (Mulford, 1966; Lehnert et al., 1968). Other chelating agents that can be used to extract cadmium into an organic solvent are dithiozone (Saltzman, 1953) and sodium diethyl dithiocarbamate (Berman, 1967).

Ion exchange resins have also been applied for separating and concentrating cadmium from digested food samples (Baetz & Kenner, 1974) and from urine and blood samples acidified with hydrochloric acid (Lauwerys et al., 1974c; Vens & Lauwerys, 1982).

### 2.2.3 Methods for quantitative determination

A number of methods have been developed for cadmium analysis, but none of them are known to produce absolutely "true" concentrations of cadmium in any material. The accuracy of a method also depends on how high the concentration is.

The nearest approximation to the "true" value when analysing complex organic materials with low cadmium concentration is probably attained with the isotope dilution mass spectrometry (IDMS) method carried out in "ultraclean" facilities. However, IDMS is extremely expensive com-pared with other methods, and has been used mainly for quality control of other methods and for certified reference materials.

The most commonly used methods, at present, are atomic absorption spectrometry, electrochemical methods, and neutron activation analysis. These three methods will be discussed in detail below. Other methods are colorimetry with dithiozone, atomic emission spectrometry, atomic fluorescence spectrometry, and proton-induced X-ray emissions (PIXE) analysis. Analytical methods for cadmium have been reviewed by Friberg et al. (1986).

In addition, *in vivo* analysis of cadmium in kidney and liver has been carried out by certain investigators (Ellis et al. 1981a; Roels et al. 1981b; Roels et al. 1983a, 1983b). The method uses the principles of neutron activation and is discussed in section 8.2.1.6 of this monograph.

The validity and accuracy of any method should ideally be ascertained by adequate quality assurance data (section 2.3). In the absence of such data, the results should at least be accompanied by intra-laboratory quality control data, results of analysis of certified standard materials, or inter-laboratory comparison data (section 2.3). Older basic chemical analysis methods may be as accurate as newer more complex and expensive methods, at least in the higher concentration range, and no analytical results should be dismissed or accepted until the method used has been carefully evaluated.

#### 2.2.3.1 *Atomic absorption spectrometry*

The basic principle is to pass the sample into a high-temperature flame (burner) or furnace and measure the absorption from the atoms in the ground state. A lamp with a cathode made up from the pure metal or an alloy of the desired element, emitting the narrow line spectrum of this element, is used as an external light source. Atomic absorption spectrometry (AAS) is the method most commonly used at present for cadmium determination, because the procedure is relatively simple and fast, and its detection limit is sufficient for most environmental and biological materials. The absorption is measured at the cadmium line (228.8 nm).

There are two main methods for atomization of a sample, the flame method and electrothermal atomization (ETA). The latter is also called the heated graphite atomization, graphite furnace or flameless method. Flame methods are generally used for liquid samples that can be aspirated into a flame, usually an air-acetylene flame. The detection limit for cadmium in pure water is of the order of 1-5 mg/litre and, in biological materials, it is about 0.1 mg/kg. At lower levels, it is usually necessary to increase the sensitivity by some accessory or by preconcentration during sample treatment. One important modification of the flame technique is the use of a micro-crucible or cup made of nickel (Delves, 1970; Fernandez & Kahn, 1971; Ediger & Coleman, 1973). The atoms are held much longer in the light beam that passes through the tube, and this increases the sensitivity considerably.

ETA methods have undergone rapid development in recent years. The sample, usually in solution (1-100 ml), is first inserted into a graphite furnace, which is surrounded by a constant flow of inert gas, such as argon or nitrogen. The temperature is then increased in order to dry, ash, and atomize the sample. During atomizing, the specific absorption from cadmium is deduced from the light beams passing through or just above the furnace. The detection limit is extremely low (of the order of a few pg). There have been several detailed reports describing the analysis, using ETA, of cadmium in biological samples such as blood and urine (Lundgren, 1976; Castilho & Herber, 1977; Stoeppler & Brandt, 1978, 1980; Vesterberg & Wrangskogh, 1978; Gardiner et al., 1979; Delves & Woodward, 1981; Subramanian & Meranger, 1981; Jawaid et al., 1983). The lowest detectable concentration of cadmium in blood and urine using ETA is of the order of 0.1-0.3 mg/litre (Delves, 1982).

Although the atomic absorption spectrometry for cadmium is specific, the method is not free from problems when applied to measurements in biological samples. Several important sources of interference exist, especially light scattering from particles and nonspecific absorption from the broad molecular absorption band formed by, for instance, sodium chloride and phosphate ions. Piscator (1971) showed that sodium chloride, at a concentration of 0.5 mol/litre, gave a signal corresponding to a concentration of 0.1 mg cadmium/litre when using ordinary air-acetylene flame atomic absorption equipment without background correction. The actual concentration was less than 0.4 mg cadmium/litre. Many problems related to interfering salts may be compensated by the use of a background correction system. A deuterium or hydrogen lamp is usually used. The nonspecific absorption can thus be measured and the signal, measured as the difference between the specific and nonspecific absorption, is proportional to the actual cadmium concentration (Kahn & Manning, 1972). Background correction for fine structure nonspecific absorption can also be made by utilizing the Zeeman effect on incoming light when it is modulated by strong magnetic fields (Koizumi et al., 1977; Alt, 1981; Pleban et al., 1981). Some kind of background correction is necessary when the microcrucible or electrothermal atomization techniques are used for cadmium analysis, since the nonspecific absorption increases as the atoms are kept in the light for a relatively long period of time.

#### 2.2.3.2 Electrochemical methods

Cadmium can be determined by different types of electrochemical methods such as classic polarographic methods or the more recently developed anodic stripping voltammetry and cadmium-selective electrodes. The basic principle behind the electrochemical methods is the change in the electrochemical potentials formed when electrons are transferred from one metal to another. A dropping mercury electrode is placed in a solution where the metal concentration is to be determined. By changing the charge of the electrode, different metals will be reduced and form an amalgam (a solid solution of metal atoms and mercury) with the mercury electrode. Polarographic waves can thus be recorded. Different metals can be determined simultaneously in a liquid sample, since they form amalgams at different charges.

Anodic stripping voltammetry is based on the reverse process, i.e. the release of metals that have already been reduced and are bound to the mercury electrode. During oxidation and release from the amalgam, a peak current can be recorded at a potential that is characteristic for the particular metal. Anodic stripping voltammetry is one of the most sensitive methods for cadmium determination available. The most crucial aspects are complete destruction of all organic materials and the transfer of cadmium ions from the sample into a non-contaminated electrolyte. The method is especially suitable for water analysis, where no sample treatment is necessary (Piscator & Vouk, 1979), but has also been used for the measurement of cadmium in various biological materials such as urine (Jagner et al., 1981), foodstuffs, and tissues (Danielsson et al., 1981). In urine, a detection limit of about 0.1 mg/litre was obtained when using a computerized potentiometric stripping analysis (Jagner et al., 1981).

Specific cadmium-selective electrodes are commercially available, but their sensitivity is insufficient for cadmium measurement in most biological materials. Furthermore, the electrodes are not ion specific, and problems can easily arise from various contaminants in the solution used (Hislop, 1980).

### 2.2.3.3 *Activation analysis*

Cadmium has a number of stable isotopes. Irradiation with neutrons yields new radioactive cadmium isotopes, which can be quantitatively measured on the basis of their specific energy and half-life. A procedure for determining cadmium in human liver samples by neutron activation analysis has been reported by Halvorsen & Steinnes (1975). The irradiated sample is usually digested before the radioactivity is measured. Sometimes, it may be necessary to concentrate cadmium by chemical methods and to separate the cadmium ions from other isotopes that have an energy spectrum overlapping the one for cadmium before measurement can be carried out. Non-radioactive cadmium can also be added after irradiation to enable measurement of the recovery after digestion and various concentration steps. The detection limit for neutron activation analysis is low, of the order of 0.1-1 mg cadmium/kg or 0.1-1 mg/litre, in most biological materials. However, the method is expensive since the samples have to be irradiated in a reactor, and so it is not normally used for screening programmes. Neutron activation analysis has been used as a reference method for accuracy tests of other methods (Kjellström et al., 1975b; Kjellström, 1979; Jawaid et al., 1983).

Neutron activation analysis is not ideal for liquid samples such as blood and urine, where the detection limit of the method is very close to the normal values. Furthermore, ampoules filled with liquid samples sometimes explode as gases are formed when the sample is irradiated in the reactor.

Irradiation with protons, proton-induced X-ray emission (PIXE), can also be used for activation analysis of cadmium. Several elements are measured at the same time. The main advantage of the method is its ability to detect and quantify cadmium in very small samples such as thin slices of tissues weighing less than 1 mg (Hasselmann et al., 1977; Mangelson et al., 1979).

### 2.2.3.4 In vivo *methods*

A non-invasive technique for *in vivo* determination of liver and kidney cadmium has been developed (Biggin et al., 1974; Harvey et al., 1975; McLellan et al., 1975) using the principle of neutron activation analysis and taking advantage of the very large capture cross-section area for thermal neutrons of one of the naturally-occurring stable isotopes of cadmium ($^{113}$Cd; natural abundance, 12.26%). A portable system using a $^{238}$Pu-Be source of neutrons (instead of the original, which was cyclotron dependent) has made this technique more easily available (Thomas et al., 1976).

The lowest detection limit for "field-work" techniques currently in use for this method is about 1.5 mg/kg in liver and 15 mg/kg in whole kidney (Ellis et al., 1981a). These limits are too high to measure accurately tissue levels in people with "normal" environmental exposure (section 6.4). In people with occupational exposure, cadmium levels of up to 100 mg/kg in liver and 400 mg/kg in whole kidney have been reported (Ellis et al., 1981a; Roels et al., 1981b). The method is still not developed to its full capacity, and the results are greatly affected by, for instance, the variability in the location of the kidney (Al-Haddad et al., 1981).

An alternative method for *in vivo* determination of cadmium concentration in kidney cortex using X-ray-generated atomic fluorescence (XRF method) has been reported (Ahlgren & Mattson, 1981; Christofferson & Mattson, 1983). Skerfving et al. (1987) found the limit of detection to be 17 $\mu$g/g kidney cortex (three standard deviations above the background). The precision is 23%.

The validity and accuracy of these *in vivo* neutron activation and XRF methods have not been studied sufficiently. A comparison of the results obtained by *in situ* determination of liver and kidney cadmium in deceased people with those found by chemical analysis of the same tissues is needed.

## 2.3 Quality control and quality assurance

### 2.3.1 Principles and need for quality control

There is a great need for strict quality control procedures in the monitoring of trace elements in biological materials. The purpose of these is to ensure that published data are as accurate as possible. Quality control involves intra-laboratory or inter-laboratory procedures that check whether the method gives acceptable results on samples with known concentrations. Quality assurance is usually given a broader meaning to cover the whole system of activities that are carried out to increase the quality of the operation. Thus, quality assurance includes not only the chemical analysis, but also the whole pre-analytical chain, data, handling, reporting, etc.

A review of published data (Vahter, 1982) showed that mean blood cadmium concentrations in the general population as high as 20-50 mg/litre have been reported. Such values are definitely unrealistic (section 6.2). Furthermore, most published reports lack quality control or quality assurance data. Valid comparisons of cadmium exposure based on blood cadmium levels can, therefore, seldom be made. Results from interlaboratory comparisons amplify the need for quality control (section 2.3.3).

### 2.3.2 Comparison of methods and laboratories

As indicated above, AAS (direct or combined with a separation procedure by organic solvent extraction or ion exchange) is the common method and can be applied to ordinary environmental or biological samples. Each of the other methods has its particular characteristics and can be used effectively according to the need for sensitivity and to the type of sample. Of special concern are methods used for the determination of cadmium in, for instance, food, blood, and urine, where cadmium concentrations are generally low and the matrices are complicated. Attempts to evaluate the accuracy by comparing the proposed method with another method have seldom been made. When testing a new method for the determination of cadmium or a new application of a method to a different type of sample, it is advisable to compare it with another method based on quite different principles.

Since the principle of neutron activation analysis is quite different from that of other methods, it is a good method for comparison. Thus, Linnman et al. (1973) and Kjellström et al. (1974, 1975b) found good agreement between a flameless atomic absorption method and destructive neutron activation analysis (the sample is irradiated and then treated chemically so the original material is "destroyed") for cadmium in wheat at concentrations down to around 20 mg/kg wheat. In the latter study (Kjellström et al., 1975b), good agreement was also found between cadmium concentrations in urine (above 5 μg/litre), determined by AAS after extraction into organic solvent, and cadmium concentrations determined by neutron activation. Because of technical problems of neutron activation analysis of liquid samples, Kjellström et al. (1975b) could not evaluate the accuracy at urine concentrations of around 1 μg/litre. However, Jawaid et al. (1983) have used neutron activation to confirm the accuracy of atomic absorption analyses of urine in the range of 0.2-4 μg/litre.

Further comparisons of destructive neutron activation analysis and different AAS methods conducted in different laboratories have been carried out for faeces, rice, wheat, liver, and muscle (Kjellström, 1979). The best agreement was found for liver, in which the cadmium concentrations were the highest, but, there was also reasonable agreement between most of the methods in the case of other materials.

Another possibility for testing a method is to add radioactive cadmium to the samples (Kjellström et al., 1974) or to inject radioactive cadmium into animals and then compare results of radioactive measurements with those obtained by chemical analysis.

Since there has been a need for comparing cadmium levels in different areas of the world, studies among laboratories in different countries have been undertaken to ensure that the analytical methods give comparable results.

An intercomparison programme involving several European laboratories, which used flame atomic absorption, flameless atomic absorption, colorimetry, polarography, and anodic stripping voltammetry, indicated great variability in results (Lauwerys et al., 1975). Thus, reported concentrations in the same sample of blood were from 1 to 92 μg/litre in one case, from 0 to 73 μg/litre in another, and from 0 to 110 mg/litre in a third. A wide range of values was also reported in the case of aqueous solutions. Only 29% of participating laboratories measured cadmium in blood with sufficient precision. The conclusion from this study was that several participating laboratories had not yet adequately developed the technique required for precisely measuring cadmium in blood, urine, and water.

### 2.3.3 Quality assurance

An extensive quality assurance programme of cadmium analysis involving laboratories in nine different countries has been carried out (Vahter, 1982). This was a part of the UNEP/WHO Global Environmental Monitoring Programme and involved the analysis of cadmium in blood and kidney tissue as well as of lead in blood. A series of quality control samples (spiked specimens), the concentrations being known or unknown to the participating laboratories, was used to check the accuracy of the methods before the population samples were analysed. This procedure was repeated up to 12 times, development work on the methods being carried out in between, in order to improve the accuracy of the methods. After improvement of the techniques and practice, the agreement became excellent. An overview of various aspects of quality assurance has been presented by Friberg (1988).

## 2.4 Conclusions

There are several methods available for the determination of cadmium in biological materials. Atomic absorption spectrometry (AAS) is the most widely used, but careful treatment of samples and correction for interference is needed for the analysis of samples with low cadmium concentrations. It is strongly recommended to accompany analysis with a quality assurance programme. At present, it is possible under ideal circumstances to determine concentrations of about 0.1 μg/litre in urine and blood and 1-10 μg/kg in food and tissue samples.

# 3. SOURCES OF HUMAN AND ENVIRONMENTAL EXPOSURE

The metal cadmium belongs, together with copper and zinc, to group IIb of the Periodic Table. It is a relatively rare element and is not found in the pure state in nature. Cadmium is mainly associated with the sulfide ores of zinc, lead, and copper, although purification first took place in 1817 from zinc carbonate. Commercial production only became significant at the beginning of this century. Cadmium is often considered as a metal of the 20th century; indeed, over 65% of the cumulative world production has taken place in the last two decades (Wilson, 1988).

Cadmium is commonly regarded as a pollutant of worldwide concern. The metal has been reviewed by the International Register of Potentially Toxic Chemicals of the United Nations Environment Programme. As a result, it has been included on the list of chemical substances and processes considered to be potentially dangerous at the global level (IRPTC, 1987).

## 3.1 Natural occurrence and cycling

Cadmium is widely distributed in the earth's crust at an average concentration of about 0.1 mg/kg. However, higher levels may accumulate in sedimentary rocks, and marine phosphates often contain about 15 mg cadmium/kg (GESAMP, 1984). Weathering also results in the riverine transport of large quantities of cadmium to the world's oceans and this represents a major flux of the global cadmium cycle; an annual gross input of 15 000 tonnes has recently been estimated (GESAMP, 1987).

Some black shale deposits in parts of the United Kingdom and USA contain elevated cadmium levels, thus leading to high soil concentrations in these areas (Lund et al., 1981). High soil concentrations are more commonly found in areas containing deposits of zinc, lead, and copper ores. Indeed, such areas are often characterized by both soil and aquatic contamination at the local level. The mining of these ore bodies has further increased the extent of such contamination. In background areas away from such deposits, surface soil concentrations of cadmium typically range between 0.1 and 0.4 mg/kg (Page et al., 1981) while fresh water contains < 0.01-0.06 ng/litre (Shiller & Boyle, 1987).

Volcanic activity is a major natural source of cadmium release to the atmosphere. Emissions of cadmium take place both during episodic eruptions and continuous low-level activity. Difficulties exist in quantifying the global flux from this source but an estimate of 100-500 tonnes (Nriagu, 1979) has been made. Deep sea volcanism is also a source of environmental cadmium release, but the role of this process in the global cadmium cycle remains to be quantified.

Older measurements of cadmium in the atmosphere and marine waters from background areas generally yielded much higher values than those obtained by more recent studies. Improved sampling and analytical techniques are considered to be responsible for these changes. Recent measurements of atmospheric concentrations in remote areas are typically in the range of 0.01-0.04 $ng/m^3$ (GESAMP, 1985). Airborne cadmium concentrations around volcanoes can be markedly elevated; for example, the plume of Mount Etna, Sicily, contains about 90 $ng/m^3$ (Buatmenard & Arnold, 1978).

Current measurements of dissolved cadmium in surface waters of the open oceans give values of $< 5$ ng/litre. The vertical distribution of dissolved cadmium in ocean waters is characterized by a surface depletion and deep water enrichment, which corresponds to the pattern of nutrient concentrations in these areas (Boyle et al., 1976). This distribution is considered to result from the absorption of cadmium by phytoplankton in surface waters, its transport to the depths incorporated in biological debris, and its subsequent release. In contrast, cadmium is enriched in the surface waters of areas of upwelling, and this leads to elevated levels in plankton unconnected with human activity (Martin & Broenkow, 1975; Boyle et al., 1976). Oceanic sediments underlying these areas of high productivity can contain markedly elevated cadmium levels as a result of inputs associated with biological debris (Simpson, 1981).

Ice and snow deposits from the polar regions represent a unique historical record of pollutants in atmospheric precipitation. However, the problems of contamination are great and no reliable data are at present available from historic samples; this prevents an insight into temporal changes in the cycling of cadmium. Nevertheless, current ice samples have been analysed; those from the Arctic contain on average 5 pg/g, while corresponding values from the Antarctic (0.3 pg/g) are much lower (Wolff & Peel, 1985).

## 3.2 Production

Cadmium is a by-product of zinc production. As a result, the level of cadmium output has closely followed the pattern of zinc production, little being produced prior to the early 1920s. The subsequent rapid increase corresponded to the commercial development of cadmium electroplating. Worldwide production reached a plateau in the 1970s but in the 1980s output appeared to be increasing again (Wilson, 1988b). The worldwide production of cadmium in 1987 was 18 566 metric tonnes (Wilson, 1988b).

## 3.3 Uses

Cadmium has a limited number of applications but within this range the metal is used in a large variety of consumer and industrial materials. The principal applications of cadmium fall into five categories: protective plating on steel; stabilizers for polyvinyl chloride (PVC); pigments in plastics and glasses; electrode material in nickel-cadmium batteries; and as a component of various alloys. Detailed consumption statistics are only available for a limited number of countries but from these it is apparent that the pattern of use can vary considerably from country to country (Wilson, 1988b).

Examination of the reported trends in cadmium consumption over the last 25 years reveals considerable changes in the relative importance of the major applications. The use of cadmium for electroplating represents the most striking decrease; in 1960 this sector accounted for over half the cadmium consumed worldwide, but in 1985 its share was less than 25% (Wilson, 1988b). This decline is usually linked to the widespread introduction of progressively stringent effluent limits from plating works and, more recently, to the introduction of general restrictions on cadmium consumption in certain countries. In contrast, the use of cadmium in batteries has shown considerable growth in recent years from only 8% of the total market in 1970 to 37% by 1985. The use of cadmium in batteries is particularly important in Japan and represented over 75% of the total consumption in 1985 (Wilson, 1988b).

Of the remaining applications of cadmium, pigments and stabilizers are the most important, accounting for 22% and 12%, respectively, of the world total in 1985. The share of the market by cadmium pigments remained relatively stable between 1970 and 1985 but the use of the metal in stabilizers during this period showed a considerable decline, largely as a result of economic factors. The use of cadmium as a constituent of alloys is relatively small and has also declined in importance in recent years, accounting for about 4% of total cadmium use in 1985 (Wilson, 1988b).

## 3.4 Sources of environmental exposure

Numerous human activities result in the release of significant quantities of cadmium to the environment. The relative importance of individual sources varies considerably from country to country. The major sources of anthropogenic cadmium release can be divided into three categories. The first is made up of those activities involved in the mining, production, and consumption of cadmium and other non-ferrous metals. The second category consists of inadvertent sources where the metal is a natural constituent of the material being processed or consumed. Sources associated with the disposal of materials that had earlier received cadmium discharges or discarded cadmium products make up the third category.

### 3.4.1 *Sources of atmospheric cadmium*

Estimates of cadmium emissions to the atmosphere from human and natural sources have been carried out at the worldwide, regional, and national levels; examples of such inventories are shown in Table 1.

The median global total emission of the metal from human sources in 1983 was 7570 tonnes (Nriagu & Pacyna, 1988) and represented about half the total quantity of cadmium produced in that year. In comparison, the worldwide emission of lead from human activities was about 10% of the total lead produced in 1983 (Nriagu & Pacyna 1988). In both the European Economic Community (EEC) and on a worldwide scale (Nriagu, 1979), about 10-15% of total airborne cadmium emissions arise from natural processes, the major source being volcanic action.

Considerable differences exist in the relative importance of different sources of atmospheric cadmium between the worldwide situation and that in the United Kingdom and the EEC as a whole. This is particularly marked for non-ferrous metal production, which accounts for about 75% of the total anthropogenic emissions worldwide but only 25% in the EEC. This partly reflects the extensive emission controls operated by these industries in Europe compared with many parts of the world. In addition, of the two

Table 1. Estimates of atmospheric cadmium emissions (tonnes/year) from human activities on a national, regional and worldwide basis

| Source | United Kingdom[a] | EEC[b] | Worldwide[c] |
|---|---|---|---|
| Natural sources | ND | 20 | 800[d] |
| Non-ferrous metal production | | | |
| mining | ND | ND | 0.6-3 |
| zinc and cadmium | | 20 | 920-4600 |
| copper | 3.7 | 6 | 1700-3400 |
| lead | | 7 | 39-195 |
| Secondary production | | ND | 2.3-3.6 |
| Production of cadmium-containing substances | ND | 3 | ND |
| Iron and steel production | 2.3 | 34 | 28-284 |
| Fossil fuel combustion | | | |
| coal | 1.9 | 6 | 176-882 |
| oil | | 0.5 | 41-246 |
| Refuse incineration | 5 | 31 | 56-1400 |
| Sewage sludge incineration | 0.2 | 2 | 3-36 |
| Phosphate fertilizer manufacture | ND | ND | 68-274 |
| Cement manufacture | 1 | ND | 8.9-534 |
| Wood combustion | ND | ND | 60-180 |
| TOTAL EMISSIONS | 14 | 130 | 3900-12 800 |

[a] From: Hutton & Symon (1986); data apply to 1982-1983
[b] From: Hutton (1983); data apply to 1979-1980 (the EEC consisted, at that time, of Belgium, Denmark, Federal Republic of Germany, Italy, Luxembourg, The Netherlands, Republic of Ireland, and the United Kingdom)
[c] From: Nriagu & Pacyna (1988); data apply to 1983
[d] From: Nriagu (1979)
ND Not determined

basic methods of zinc production, thermal smelting and electrolyte refining, only the former releases significant atmospheric cadmium emissions. In recent years, electrolytic refining has assumed the major share of the world's production of zinc and cadmium and has largely replaced thermal processes in Europe. The once important vertical and horizontal retort smelters, which emit large quantities of atmospheric cadmium, have been phased out in most developed countries, but are still in operation in several developing countries (ILZSG, 1988).

Other industries that employ thermal processes, e.g., iron production, fossil fuel combustion, and cement manufacture, all release airborne cadmium, the metal being a natural constituent of the raw materials. The cadmium content of these materials is generally relatively low but this is offset by the vast quantities consumed. Furthermore, in common with other thermal processes, the elevated temperatures employed result in the volatilization of cadmium. It subsequently condenses in a preferential manner on the smallest particles in the stack gases, the size range least efficiently retained by conventional particulate control measures (Smith, 1982). Despite mechanisms that enhance the release of cadmium, the quantities emitted from the three processes are now considered to be smaller than they were in the past, particularly in the case of fossil fuel combustion (Rauhut, 1980). Municipal refuse is a waste-related source, the cadmium being derived from discarded nickel-cadmium batteries and plastics that contain cadmium pigments and stabilizers. The incineration of refuse, a practice generally restricted to developed countries, is a major source of atmospheric cadmium release at the national, regional, and worldwide levels (Table 1). Indeed, this activity accounts for about one third of the total cadmium emissions in the United Kingdom and the EEC as a whole. Cadmium release from this sector originates from a large number of plants, while the emissions from the non-ferrous metal industry are derived from relatively few facilities.

Sewage sludge receives cadmium from industrial sources, particularly from the discharges of plating operations and pigment works. One disposal option, the incineration of sewage sludge, is a relatively minor source of airborne cadmium, reflecting the small quantities of sludge disposed of in this manner (Table 1).

Steel production can also be considered as a waste-related source, as large quantities of cadmium-plated steel scrap are recycled by this industry, at least in developed countries. As a result, steel production is responsible for considerable emissions of atmospheric cadmium.

### 3.4.2 Sources of aquatic cadmium

Non-ferrous metal mines represent a major source of cadmium release to the aquatic environment. Contamination can arise from mine drainage water, waste water from the processing of ores, overflow from the tailings pond, and rainwater run-off from the general mine area. The release of these effluents to local watercourses can lead to extensive contamination downstream of the mining operation. The cadmium content of the ore body and mine management policies, as well as climatic and geographical conditions, all influence the quantities of cadmium released from individual sites. Flood and storm conditions, for example, will enhance the mobilization of cadmium contained in particulate material. Aquatic inputs of cadmium are not restricted to active mine sites, and mines disused for many years can still be responsible for the continuing contamination of adjacent watercourses (Johnson & Eaton, 1980).

At the global level, the smelting of non-ferrous metal ores has been estimated to be the largest human source of cadmium release to the aquatic environment (Nriagu & Pacyna, 1988). Discharges to fresh and coastal waters arise from liquid effluents produced by gas scrubbing together with the site drainage waters.

Concerning the locations where environmental health effects of cadmium have been reported, the water and air contamination from non-ferrous metal mining and production are the predominant sources of cadmium. All the major areas of Japan with elevated cadmium levels have been affected by these sources (Tsuchiya, 1978), although contamination through the natural mobilization of cadmium from ore bodies may also have been involved.

Cadmium is a natural constituent of rock phosphates and deposits from some regions of the world contain markedly elevated levels of the metal. The manufacture of phosphate fertilizer results in a redistribution of the cadmium in the rock phosphate between the phosphoric acid product and the gypsum waste. In many cases, the gypsum is disposed of by dumping in coastal waters, which leads to considerable cadmium inputs. Some countries, however, recover the gypsum for use as a construction material and thus have negligible cadmium discharges (Hutton, 1982).

The atmospheric fall-out of cadmium to fresh and marine waters represents a major input of cadmium at the global level (Nriagu & Pacyna, 1988). Indeed, a GESAMP study of the Mediterranean Sea indicated that this source is comparable in magnitude to the total river inputs of cadmium to the region (GESAMP, 1985). Similarly, large cadmium inputs to the North Sea (110-430 tonnes/year) have been estimated, based on the extrapolation from measurements of cadmium deposition along the coast (van Aalst et al., 1983a,b). However, another approach based on model simulation yielded a modest annual input of 14 tonnes (Krell & Roeckner, 1988).

Acidification of soils and lakes may result in enhanced mobilization of cadmium from soils and sediments and lead to increased levels in surface and ground waters (WHO, 1986). The corrosion of soldered joints or zinc galvanized plumbing by acidic waters can dissolve cadmium and produce increased levels of the metal in drinking-water. In one study from Sweden, cadmium levels in tap water from areas susceptible to acidic deposition were double those from a control area (Svensson et al., 1987).

### 3.4.3 Sources of terrestrial cadmium

Solid wastes from a variety of human activities are disposed of in landfill sites, resulting in large cadmium inputs at the national and regional levels when expressed as a total tonnage (Hutton, 1982; Hutton & Symon, 1986). However, this simple approach exaggerates the significance of landfilled cadmium in certain high volume wastes with relatively low concentrations of cadmium. Examples include the ashes from fossil fuel combustion, waste from cement manufacture, and the disposal of municipal refuse and sewage sludge. Of greater potential environmental significance are the solid wastes from both non-ferrous metal production and from the manufacture of cadmium-containing articles, as well as the ash residues from refuse incineration. All three waste materials are characterized by elevated cadmium levels and as such require disposal to controlled sites to prevent the mobilization of the cadmium in ground water.

Soil cadmium contamination is a characteristic feature around non-ferrous metal mines and smelters, particularly in the case of those handling zinc ores. Contamination from mining is generally local but may be widespread in areas of high mineral content (Tsuchiya, 1978). Soil contamination from smelters is generally greatest next to the source and decreases exponentially with distance, although cadmium concentrations can still be above the background level 20 km from the source (Buchauer, 1972). Shipham, United Kingdom, is a site of extreme soil cadmium contamination. Between 1650 and 1850 the village of Shipham was the site of a major zinc mine. Once the mining stopped the area was flattened and developed for agriculture and housing. Cadmium levels in agricultural and garden soils are some of the highest ever reported worldwide (Thornton, 1988).

The agricultural application of phosphate fertilizers represents a direct input of cadmium to arable soils. The cadmium content of phosphate fertilizers varies widely and depends on the origin of the rock phosphate. It has been estimated that fertilizers of West African origin contain 160-255 g cadmium/tonne of phosphorus pentoxide, while those derived from the southeastern USA contain only 35 g/tonne (Hutton, 1982).

The annual rate of cadmium input to arable land from phosphate fertilizers had been estimated for the countries of the EEC, taking into account differences in application rates and the cadmium contents of the fertilizers used (Hutton, 1982). The average cadmium input (5 g/ha) only represents about 1% of the surface soil cadmium burden. Despite the relatively small size of this input, long-term continuous application of phosphate fertilizers has been shown to cause increased soil cadmium concentrations (Williams & David, 1973, 1976; Andersson & Hahlin, 1981).

The application of municipal sewage sludge to agricultural soil as a fertilizer can also be a significant source of cadmium. In many industrialized countries, control measures have reduced the cadmium content of sewage sludge and at the same time national and regional regulations have limited the input of cadmium from agricultural sludge applications (Davis, 1984). Nevertheless, large increases in soil cadmium concentration have resulted in the past from the application of contaminated sludge in both North America and Europe (Davis, 1984). Even today, the high application rates used for sewage sludge result in relatively large cadmium inputs, a value of 80 g/ha having been estimated for the United Kingdom (Hutton & Symon, 1986). On a national or regional basis, however, these inputs are much smaller than those from either phosphate fertilizers or atmospheric deposition (see section 4.2).

## 3.5 Conclusions

Cadmium is a relatively rare element and current analytical procedures indicate much lower concentrations of the metal in environmental media than do older measurements. At present, it is not possible to determine whether human activities have caused a historic increase in cadmium levels in the polar ice caps.

Commercial cadmium production started at the beginning of this century. The pattern of cadmium consumption has changed in recent years with significant decreases in electroplating and increases in batteries and specialized electronic uses. Most of the major uses of cadmium employ it in the form of compounds that are present at low concentration. This makes it difficult to recycle cadmium in order to decrease the potential for environmental contamination. Restrictions on certain uses of cadmium imposed by a few countries may have widespread impact on the applications of cadmium.

Cadmium is released to the air, land, and water by human activities. In general, the two major sources of contamination are the production and consumption of cadmium and other non-ferrous metals and the disposal of wastes containing cadmium. Areas in the vicinity of non-ferrous mines and smelters often show pronounced cadmium contamination.

# 4. ENVIRONMENTAL TRANSPORT, DISTRIBUTION, AND TRANSFORMATION

## 4.1 Atmospheric deposition

Cadmium is removed from the atmosphere by dry deposition and by precipitation. Total deposition rates have been measured at numerous localities worldwide and values have generally been found to increase in the order: background < rural < urban < industrial. In rural areas of Scandinavia, annual deposition rates ranged from 0.4 to 0.9 g/ha (Laamanen, 1972; Andersson, 1977). Similarly, in a rural region of Tennessee, USA, a deposition rate of 0.9 g/ha was observed (Lindberg et al., 1982). Hutton (1982) concluded that 3 g/ha per year is a representative value for the atmospheric deposition of cadmium to agricultural soils in rural areas of the EEC. This may be compared with a corresponding input of 5 g/ha per year for these areas from the application of phosphate fertilizers (see 3.4).

Many industrial sources of cadmium possess tall stacks, which bring about the wide dispersion and dilution of particulate emissions. Indeed, it is often assumed that < 10% of such emissions are deposited locally, the remainder being available for long-range transport (Krell & Roeckner, 1988). Nevertheless, cadmium deposition rates around smelter facilities are often markedly elevated nearest the source and generally decrease rapidly with distance (Hirata, 1981). This pattern of contamination can be reflected in surface soils and vegetation, and in the former case, contamination will reflect the long-term history of metal inputs from the atmosphere. As a result, soil cadmium concentrations in excess of 100 mg/kg are commonly encountered close to long-established smelters (Buchauer, 1972). In some urban areas, the high density of non-ferrous metal works results in a city-wide elevation of cadmium deposition (Roels et al., 1981a).

The possibility that cadmium deposition is enhanced around atmospheric sources of cadmium other than smelters has been investigated on a number of occasions. One assessment of studies conducted around coal-fired power stations concluded that this source was unlikely to cause any marked local accumulation of cadmium (Chadwick & Lindman, 1982). In contrast, significant cadmium contamination was found in surface soil downwind of a phosphate fertilizer processing plant in the USA, the levels being up to 40 mg/kg (Hutchison et al., 1979). Little attention has been paid to the pattern of cadmium deposition around refuse incinerators; one study of a large facility in the United Kingdom observed moderately elevated deposition rates downwind of the plant (Hutton et al., 1988).

Crop plants growing near to atmospheric sources of cadmium may contain elevated cadmium levels (Carvalho et al., 1986). However, it is not always possible to distinguish whether the cadmium is derived directly from surface deposition or originates from root uptake, since soil levels in such areas are generally higher than normal. One study in Denmark has suggested that atmospheric deposition can also be an important direct source of cadmium in crop plants even in background areas (Hovmand et al., 1983).

## 4.2 Transport from water to soil

Rivers contaminated with cadmium can contaminate surrounding land, either through irrigation for agricultural purposes, by the dumping of dredged sediments, or through flooding (Forstner, 1980; Sangster et al., 1984). For example, agricultural land adjacent to the Neckar River, Germany, received dredged sediments to improve the soil, a practice that produced soil cadmium concentrations in excess of 70 mg/kg (Forstner, 1980).

Much of the cadmium entering fresh waters from industrial sources is rapidly absorbed by particulate matter, where it may settle out or remain suspended, depending on local conditions. This can result in low concentrations of dissolved cadmium even in rivers that receive and transport large quantities of the metal (Yamagata & Shigematsu, 1970). Rivers can transport cadmium considerable distances from the source of the input. In Japan, there are several areas where soils have been contaminated with irrigation water up to 50 km from the source (Tsuchiya, 1978).

## 4.3 Uptake from soil by plants

It has been shown repeatedly that an increase in soil cadmium content results in an increased plant uptake of the metal. This has been demonstrated for soils with naturally elevated cadmium levels (Lund et al., 1981), those contaminated by non-ferrous metal mining (Alloway et al., 1988), and those that have received cadmium via sewage sludge applications (Davis & Coker, 1980). It is this basic relationship that makes the soil-crop pathway of human exposure susceptible to increased levels of soil cadmium. Indeed, since the major sources of cadmium exposure for the general population are food and tobacco (see section 5), it is important to assess those soil and plant factors that influence cadmium uptake by crop plants.

The most important soil factors influencing plant cadmium accumulation are soil pH and cadmium concentration (Davis & Coker, 1980; Page et al., 1981). Soil cadmium is distributed between a number of pools or fractions, of which only the cadmium in soil solution is thought to be directly available for uptake by plants. Soil pH is the principal factor governing the concentration of cadmium in the soil solution. Cadmium absorption to soil particles is greater in neutral or alkaline soils than in acidic ones and this leads to increased cadmium levels in the soil solution. As a consequence, plant uptake of cadmium decreases as soil pH increases.

Other soil factors that influence the distribution of cadmium between the soil and soil solution include cation exchange capacity and the contents of the hydrous oxides of manganese and iron, organic matter, and calcium carbonate. Increases in these parameters result in decreased availability of cadmium to plants owing to a reduction of the level of cadmium in the soil solution.

A comparative study of cadmium-contaminated soils from different sources illustrates the importance of the above soil factors (Alloway et al., 1988). Soils from Shipham, United Kingdom, contained the highest total cadmium levels but the soil solution concentrations were lower than in other soils. The small proportion of soluble cadmium in Shipham soils (0.04%) was related to the high pH (7.7) and high calcium carbonate and hydrous oxide content of these soils. In contrast, a paddy soil from the Junzu Valley, Japan, contained 4% soluble cadmium and possessed a low pH (5), low calcium carbonate content, and very low hydrous oxide concentration (Alloway et al., 1988).

Much attention has been paid to the plant availability of cadmium in agricultural soils to which sewage sludge has been applied. It has been observed that the repeated application of sludge to soils can alter the availability of cadmium, and although soil cadmium levels may increase, crop levels do not always reflect this increase (Page et al., 1981). The long-term availability of cadmium to plants is uncertain, availability having been reported to remain constant, decrease, or even increase with time (Tjell et al., 1983). In another study there were no clear changes in the plant availability of cadmium over a period of five years after sewage sludge was applied to the soil (Carlton-Smith, 1987).

Concern over the long-term implications of present-day cadmium inputs to European arable soils has led to modelling studies of the future cadmium exposure for the general population (Tjell et al., 1981; Hutton, 1982). It was estimated by Tjell et al. (1981) that cadmium inputs from phosphate fertilizers and atmospheric deposition will cause an annual increase of 0.6% in Danish soil cadmium levels. The corresponding increases in crop cadmium concentrations would lead to a predicted 70% increase in dietary cadmium intake 100 years hence. Similar soil and dietary cadmium increases have been predicted for the EEC as a whole, although the precise values varied according to the soil properties and crop consumption patterns employed (Hutton, 1982).

Indirect support for these forecasts was provided by an investigation of the time trends in soil and crop cadmium levels using archived samples. Jones et al. (1987) found that the cadmium content of agricultural soils from a site in the United Kingdom had increased by 27-55% since the 1850s. Trends in the cadmium concentrations of wheat grain were less clear, possibly due to confounding factors such as changes in varieties grown and altered soil properties.

## 4.4 Transfer to aquatic and terrestrial organisms

In general, cadmium concentrations in terrestrial and aquatic biota from uncontaminated localities are low, corresponding to the geochemical abundance of this metal. However, in certain situations, cadmium displays a propensity for marked bioaccumulation, a feature that has implications for human dietary exposure and may be of toxicological significance for the organisms concerned.

It appears that cadmium shows greatest mobility in certain marine ecosystems. Phytoplankton in areas of oceanic upwelling contain raised cadmium levels (Martin & Broenkow, 1975), and filter-feeding molluscs can accumulate significant concentrations of cadmium even in coastal localities that are only moderately contaminated (Bryan et al., 1980). Oysters, in particular, are well-known cadmium accumulators, levels of up to 8 mg/kg wet weight having been recorded in New Zealand (Nielsen, 1975). Certain edible crustaceans such as crab and lobster also contain relatively high cadmium concentrations, the metal being localized in the hepatopancreas or "brown meat" (Buchet et al., 1983).

Some marine birds and mammals contain remarkably elevated cadmium burdens in the kidney and liver (Martin et al., 1976; Stoneburner, 1978; Nicholson & Osborn, 1983). In the case of oceanic species, this accumulation is probably a natural process associated with the feeding habits and longevity of the organism in question. Even so, the high cadmium levels in pelagic sea-birds have been linked in one study to morphological signs of kidney damage (Nicholson & Osborn, 1983).

Terrestrial mosses and lichens display a high capacity for retention of metals deposited from the atmosphere and these plants have been used to map both local contamination from point sources and regional patterns of cadmium deposition (MARC, 1986). The fruiting bodies of some macrofungi contain remarkably high cadmium concentrations even in areas uncontaminated with cadmium (MARC, 1986). This phenomenon has implications for human dietary exposure as some accumulator species are edible.

In addition to humans, certain long-lived terrestrial mammals such as the horse and moose may possess considerable cadmium burdens in the kidney and liver (Elinder & Piscator, 1978; Frank et al., 1981; Jeffery et al., 1989). It has been shown that cadmium accumulates with age in horse kidney.

## 4.5 Conclusions

Increases in soil cadmium content result in an increase in the uptake of cadmium by plants; the pathway of human exposure from agricultural crops is thus susceptible to increases in soil cadmium. The uptake by plants from soil is greater at low soil pH. Processes that acidify soil (e.g., acid rain) may therefore increase the average cadmium concentrations in foodstuffs. The application of phosphate fertilizers and atmospheric deposition are significant sources of cadmium input to arable soils in some parts of the world; sewage sludge can also be an important source at the local level. These sources may, in the future, cause enhanced soil and hence crop cadmium levels, which in turn may lead to increases in dietary cadmium exposure. In certain areas, there is evidence of increasing cadmium content in food.

Edible free-living food organisms such as shellfish, crustaceans, and fungi are natural accumulators of cadmium. Regular consumption of these items can result in elevated human exposure. Certain marine vertebrates contain markedly elevated renal cadmium concentrations, which, although considered to be of natural origin, have been linked to signs of kidney damage in the organisms concerned.

# 5. ENVIRONMENTAL LEVELS AND HUMAN EXPOSURE

Human uptake of cadmium occurs via the inhalation of air and the ingestion of food and drinking-water. Accidental ingestion of cadmium through the contamination of foods in contact with cadmium-containing materials has occurred in the past. Accidental high-level inhalation exposure during welding operations and cadmium smelting is still a considerable hazard.

Chronic exposure to cadmium via food or workplace air is the main concern in assessing the health risks of cadmium.

## 5.1 Inhalation route of exposure

### *5.1.1 Ambient air*

Many countries carry out regular monitoring programmes for cadmium in the air. An assessment of the available data from various European countries showed that average values range from 1 to 5 ng/m$^3$ in rural areas, 5 to 15 ng/m$^3$ in urban areas, and 15 to 50 ng/m$^3$ in industrialized areas (WHO, 1987). Examination of some individual national data (Table 2) suggests that urban values are likely to occupy the lower end of the range indicated above (McInnes, 1979; RIVM, 1988).

Much higher air cadmium concentrations are found in areas close to major atmospheric sources of the metal. However, these values can fluctuate widely as a result of changing emission characteristics and weather conditions (Muskett et al., 1979).

Studies of the particle size distributions of cadmium in urban aerosols generally show that the metal is associated with particulate matter in the respirable range (Greenberg et al., 1978). The enrichment of cadmium on these smaller particles can be linked to the behaviour of the metal in thermal facilities that are sources of airborne cadmium (see section 3.4.1).

An air quality study revealed no differences between indoor and outdoor air cadmium levels when the dwellings of non-smokers were examined (Moschandreas, 1981). However, significantly higher indoor air cadmium levels were observed in those houses where smoking took place.

Table 2. Typical levels of cadmium in ambient air

| Type of area | Cadmium concentration range ($ng/m^3$) | Sampling period[b] | Reference |
|---|---|---|---|
| Remote rural | | | |
| Pacific atoll | 0.0025-0.0046 | NR | Duce et al. (1983) |
| Europe | 0.1-0.3 | NR | Heindryckx et al. (1974) |
| Atlantic | $3 \times 10^{-6}$-$6.2 \times 10^{-4}$ | NR | Duce et al. (1975) |
| Rural | | | |
| Belgium | 1[a] | 24 h | Janssens & Dams (1974) |
| Federal Republic of Germany | 0.1-1 | < 24 h | Neeb & Wahdat (1974) |
| Japan | 1-4 | 24 h | Japanese Environment Agency (1974) |
| Urban | | | |
| Belgium | 50[a] | 24 h | Janssens & Dams (1974) |
| Federal Republic of Germany | 10-150 | < 24 h | Neeb & Wahdat (1974) |
| Japan | 3-6.3 | 1 year | Japanese Environment Agency (1974) |
| Poland | 2-51 | 1 year | Just & Kelus (1971) |
| USA (New York) | 3-23 | 1 year | Kneip et al. (1970) |

[a] Mean value    [b] NR = not reported

### 5.1.2 *Air in the working environment*

Elevated air cadmium levels arise in the smelting of non-ferrous metals and in the production and processing of cadmium-containing articles. The thermal operations associated with these processes are mainly responsible for producing cadmium dusts and, if temperatures are sufficiently high, cadmium fume.

Airborne cadmium concentrations found in the occupational setting vary considerably according to the type of industry and the specific working conditions in each plant. Markedly elevated values, in the $mg/m^3$ range, were prevalent in the 1940s to 1960s (Friberg, 1950; Adams et al., 1969; Tarasenko & Vorobjeva, 1973). Considerable improvements in occupational hygiene have taken place in developed countries since then and these have led to progressive reductions in ambient levels in the workplace. Table 3 illustrates the temporal decline in air cadmium levels in a Swedish battery factory (Adamsson, 1979). The lowest values shown in Table 3 may not be typical for all occupational facilities; levels of 1-5 mg/m$^3$ were reported for one pigment plant in the mid 1970s (De Silva & Donnan, 1981).

Table 3. Average air cadmium concentrations in a Swedish cadmium battery plant[a]

| Time period | Number of observations | Cadmium concentration (µg/m$^3$) |
|---|---|---|
| 1946 | 10 | 5000 |
| 1947-1949 | 16 | 750 |
| 1950-1960 | 94 | 650 |
| 1965-1973 | 393 | 70 |
| 1973-1975 | 373 | 40 |
| 1975-1976 | 573 | 15 |

[a] From: Adamsson (1979)

In general, only total air cadmium concentrations are monitored in the working environment; factors influencing respiratory absorption, such as the speciation of cadmium and the size distribution of the collected particles, are not taken into account. In one study of workplaces with high total airborne cadmium levels, Lauwerys et al. (1974b) found, in general, that less than 25% of the total cadmium in air was in the respirable range and that this percentage decreased as the total value increased. Cadmium-containing dust particles that are too large to be delivered to the pulmonary region of the lung can enter the gastrointestinal tract by mucociliary transfer.

### 5.1.3 The smoking of tobacco

The tobacco plant naturally accumulates relatively high cadmium concentrations in its leaves. As a result, this material represents an important source of exposure for smokers. It has been reported that one cigarette contains about 1-2 µg cadmium (Friberg et al., 1974) and that about 10% of the cadmium content is inhaled when the cigarette is smoked (Elinder et al., 1983). One study has suggested that modifications in cigarette construction and the increasing popularity of filter cigarettes have reduced cadmium exposure from this source in recent years (Scherer & Barkemeyer, 1983). Regional differences exist in the cadmium concentration of cigarettes, and lower values (0.1-0.5 µg) have been found in samples from Argentina, India, and Zambia (Nwankwo et al., 1977; Elinder et al., 1983).

Biological monitoring surveys of the general population have shown that cigarette smoking can cause significant increases in the concentration of cadmium in the kidney (Lewis et al., 1972; Vahter, 1982).

Occupationally exposed workers who smoke tobacco may be subject to higher exposure levels than their non-smoking colleagues. This may be because the original content of tobacco can be considerably increased when handled during work (Piscator et al., 1976). In addition, the hand-to-mouth route of exposure may be more important in workers who are tobacco smokers (Adamsson, 1979).

## 5.2 Ingestion routes of exposure

### 5.2.1 *Levels in drinking-water*

Drinking-water generally contains low cadmium levels and a value of 1 μg/litre or less is often assumed to be a representative value in most situations (Meranger et al., 1981). Thus, cadmium exposure from drinking-water and water-based beverages is relatively unimportant compared with the dietary contribution.

In a study of drinking-water in Seattle, USA, Sharrett et al. (1982) reported a median cadmium level of 0.01 μg/litre in tap water delivered by copper pipes; the corresponding value from homes with galvanized piping was 0.25 μg/litre. Water samples left to stand in both types of piping showed increases in cadmium levels with median values of 0.06 and 0.63 μg/litre in copper and galvanized supplies, respectively. In a survey from the Netherlands, about 99% of drinking-water samples in 1982 contained less than 0.1 μg/litre (RIVM, 1988).

### 5.2.2 *Levels in food*

The cadmium content of agricultural crops varies according to species, variety cultivated and season (Davis & Coker, 1980). The results of an extensive nationwide survey of cadmium in different classes of raw agricultural crops from uncontaminated localities illustrate the range of values encountered within and between crop classes (Wolnik et al., 1983, 1985). It is evident that cadmium is a normal constituent of most foodstuffs (Tables 4 and 5).

Table 4. Cadmium concentrations in the major types of crop from various regions of the USA[a]

| Crop | Sample size | Cadmium concentration (mg/kg wet weight) | | |
| --- | --- | --- | --- | --- |
| | | Median | Minimum | Maximum |
| Rice | 166 | 0.0045 | < 0.001 | 0.23 |
| Peanuts | 320 | 0.060 | 0.010 | 0.59 |
| Soybeans | 322 | 0.041 | 0.002 | 1.11 |
| Wheat | 288 | 0.030 | < 0.0017 | 0.207 |
| Potatoes | 297 | 0.028 | 0.002 | 0.18 |
| Carrots | 207 | 0.017 | 0.002 | 0.13 |
| Onions | 230 | 0.009 | 0.001 | 0.054 |
| Lettuce | 150 | 0.017 | 0.001 | 0.160 |
| Spinach | 104 | 0.061 | 0.012 | 0.20 |
| Tomatoes | 231 | 0.014 | 0.002 | 0.048 |

[a] From: Wolnik et al. (1983, 1985).

Meat, fish, and fruit generally contain similar cadmium levels and values of 5-10 μg/kg fresh weight are representative for these food classes. Most plant-based foodstuffs contain higher cadmium concentrations and a value of 25 μg/kg fresh weight is considered representative for the staple items, cereals and root vegetables. Offal from adult animals and certain shellfish contain even higher concentrations (see section 4.4); values in excess of 50-100 μg/kg fresh weight are considered normal. Food preparation can result in cadmium losses from plant-based items. The milling of wheat grain results in a reduction of about 50% in the cadmium content of the white flour produced (Linnman et al., 1973). The washing, peeling, and cooking of vegetables can also lead to reductions in the concentrations of cadmium but, in general, these are relatively small.

The use of glazed ceramic containers to store foodstuffs can lead to significant cadmium contamination, particularly in the case of foods that are acidic liquids (Beckman et al., 1979).

Table 5. Cadmium concentrations in different food items from various European countries (values in µg/kg fresh weight)

| Food Group | United Kingdom[a] | Finland[b] | Sweden[c] | Denmark[d] | The Netherlands[e] |
|---|---|---|---|---|---|
| Bread and cereals | 20-30 | 20-40 | 31-32 | 30 | 25-35 |
| Meat | < 20-30 | < 5-5 | 2-3 | 6-30 | 10-40 |
| Offal | | | | | |
| pork kidney | 450 | 180 | 190 | | 1000 |
| pork liver | 130 | 70 | 50 | | 100 |
| Fish | < 15 | < 5-20 | 1-20 | 14 | 15 |
| Eggs | < 30 | 4 | 1 | < 10 | 2 |
| Oils and dairy products | < 20-30 | 3-20 | 1-23 | < 30 | 10-30 |
| Sugars and preserves | < 10 | < 10 | 3 | 30 | 5 |
| Fresh fruit | < 10 | < 2 | 1-2 | 11 | 5 |
| Vegetables | | | | | |
| cabbage | < 10 | 5 | 4 | 10 | |
| cauliflower | < 20 | 10 | 10 | | |
| spinach | 120 | 150 | 43 | | |
| broccoli | 10 | 10 | | | |
| legumes | < 10-30 | < 2-30 | 1-4 | 15 | |
| lettuce | < 60 | 50 | 29 | 43 | |
| potatoes | < 30 | 30 | 16 | 30 | 30 |
| carrots | < 50 | 30 | 41 | | |

[a] From: Bucke et al. (1983)
[b] From: Koivistoinen (1980)
[c] From: Jorhem et al. (1984)
[d] From: Andersen (1979)
[e] From: RIVM (1988)

Crops grown in cadmium-contaminated localities have been shown to contain elevated levels of the metal compared with normal values. The extent of enrichment depends on several factors (see section 4.3). The cadmium concentrations in selected vegetable crops grown at three contaminated sites in the United Kingdom are shown in Table 6. Highest levels were generally

Table 6. Mean cadmium concentrations (μg/kg fresh weight) in selected vegetable crops grown at three contaminated sites in the United Kingdom

| Location | Source of cadmium contamination | Cabbage | Leafy salad | Potato | Carrot | Reference |
|---|---|---|---|---|---|---|
| Shipham | zinc mine | 250[a] | 680 | 130 | 340 | Sherlock et al. (1983) |
| Walsall | atmospheric inputs from a copper refinery | 73 | 190 | 103 | 120 | Tennant (1984) |
| Heathrow | sewage sludge applications | 24 | 180 | 150 | 150 | Chumbley & Unwin (1982) |

[a] Median value

found at Shipham, where soil cadmium concentrations are markedly elevated, and the greatest increase was noted in leafy vegetables. Potato, a staple food item, showed similar values at the three locations and these were about five times greater than background.

Large scale surveys of cadmium in rice have been carried out in areas of Japan where environmental contamination was suspected (Japanese Environment Agency, 1972, 1982). The results of the earlier survey revealed that large numbers of rice samples contained elevated cadmium levels; the corresponding data from the later study indicated that decreases had occurred over the intervening ten years. More detailed investigations at specific localities have also been carried out in Japan, often as part of studies on health effects of the general population (Table 7).

### 5.2.3 *Other sources of exposure*

Young children may ingest household dust or garden soil. This habit may be a source of cadmium exposure, as has been identified for lead (Duggan et al., 1985). The representative daily intake of dust via the hands in young children is considered to be 100 mg (Lepow et al., 1974). In an extensive survey of metals in household dusts in the United Kingdom, an average cadmium level of 6.9 mg/kg was obtained from over 4500 samples (Culbard et al., 1988). These data suggest that the hand-to-mouth route is a minor source of cadmium intake (about 0.7 μg daily).

The hand-to-mouth exposure pathway may be a significant source of cadmium in areas around point sources of the metal. In the vicinity of a small lead refinery in the United Kingdom, cadmium levels in household dust were reported to be 193 mg/kg (Muskett et al., 1979). The daily ingestion of 100 mg of this dust would result in the intake of about 20 μg cadmium. Buchet et al. (1983) observed a correlation between cadmium intake from dust and the levels of blood and urinary cadmium in children from areas of Belgium subjected to air contamination. Despite markedly elevated soil cadmium levels in the gardens of Shipham, United Kingdom, Thornton (1988) found that household dust concentrations were only four times greater (at an average of 27 mg/kg) than background.

### 5.2.4 *Daily intake of cadmium from food*

Three approaches are used for estimating the daily intake of cadmium in food. The first is the total-diet collection method in

Table 7. Environmental cadmium levels in Japan: a summary of the surveys of cadmium levels in rice and health status of local populations

| Area (prefecture) | Cadmium concentration in rice (mg/kg fresh weight) | Daily cadmium intake ($\mu$g/day)[c] | Source of cadmium contamination | Number of people[a] | Health effects reported[b] | References |
|---|---|---|---|---|---|---|
| Fuchu, Toyama | 0.6-2.0 | 600 | zinc, lead, and cadmium mine and refinery | 7650 | yes | Ishizaki et al. (1969); Kato & Abe (1978) |
| Ikuno, Hyogo | 0.2-1.0 | | silver, copper, and zinc mine | 13 000 | yes | Hyogo Prefectural Gov't (1972); Tsuchiya & Nakamura (1978) |
| Tsushima, Nagasaki | 0.5-0.8 | 213-255 | lead and zinc mines | 2400 | yes | Shigematsu et al. (1975); Takabatake (1978b) |
| Kakehashi, Ishikawa | 0.2-0.8 | 160 | copper mines | 2800 | yes | Ishizaki (1972); Kawano & Kato (1978) |
| Kosaka, Akita | 0.2-0.6 | 185 | silver and copper mines | 800 | yes | Kojima et al. (1975); Shigematsu & Kawaguchi (1978) |
| Yoshino, Yamagata | 0.6 | | gold, silver, copper, and zinc mines | 8000 | NS | Uruno et al. (1975); Shigematsu & Kawaguchi (1978) |
| Annaka and Takasaki, Gunma | 0.4-0.5 | 281 | zinc refinery | 4400 | NS | Shigematsu et al. (1975); Fukushima (1978) |

Table 7 (contd).

| | | | | | | |
|---|---|---|---|---|---|---|
| Uguisuzawa, Miyagi | 0.6-0.7 | 180 | lead and zinc mines | 800 | NS | Shigematsu et al. (1975); Takabatake (1978a,b) |
| Watarase, Gunma | 0.3 | | copper mine | 4700 | NS | Fukushima et al. (1975); Fukushima (1978) |
| Shimoda, Shizuoka | 0.4-1.1 | | gold and copper mine | 1100 | NS | Tsuchiya (1978) |
| Bandai, Fukushima | 0.2-0.4 | | zinc refinery | 1800 | NS | Shigematsu & Kawaguchi (1978) |
| Kurobe, Toyama | 0.6 | | copper refinery | 8000 | NS | Tsuchiya (1978) |
| Kiyokawa, Oita | 0.2-0.5 | 391 | tin, copper, lead, zinc, and arsenic mine | 700 | NS | Takabatake (1978a); Shigematsu et al. (1975) |
| Ohmuta, Fukuoka | 0.7 | | zinc refinery | 2540 | NS | Yamamoto (1972); Tsuchiya (1978) |

[a] Indicates the approximate number of people living in exposed area. The figure usually includes only people over 30 years old considered to consume rice with more than 0.4 mg cadmium/kg.

[b] NS denotes health examinations were made, but effects were not significantly different from those in control areas.

[c] From: Tsuchiya (1978)

which the foods are prepared for consumption and are analysed either individually or combined in one or more food group composites in proportions based on available food consumption data. The total cadmium intake is calculated as the product of the concentration and the estimated amount of food eaten. In the second approach, a market basket study, representative samples of individual foodstuffs are collected from retail outlets and analysed. The cadmium concentrations are then multiplied by the average amount of intake of each foodstuff to give the cadmium intakes for each food item. The sum gives the total dietary intake. The third way of estimating cadmium intake is the collection of a duplicate sample of the meals consumed. The combined food sample is homogenized and the cadmium analysed. Table 8 presents some published estimates of dietary cadmium intakes from different countries based on these three methods.

Another method for estimating the daily intake of cadmium is to determine the daily faecal output, because only about 5% of ingested cadmium is absorbed on average (section 6.1.2). In Table 9 the available data on faecal cadmium are summarized. There is general agreement with the data presented in Table 8, but in the USA the estimated dietary exposure based on faecal analysis is considerably lower than direct estimates of dietary intake.

Tables 8 and 9 show that daily intakes of cadmium in Europe, New Zealand, and the USA are usually about 10-25 μg. These are average values, and large individual variations do occur due to variability in dietary habits and age-dependent changes in energy intake. The highest daily intake of cadmium is likely to occur among teenagers, since they have the highest caloric intake (Kjellström et al., 1978). Individuals from the general population who are extreme consumers of certain food items with elevated cadmium levels may have exposure levels above the average. It has been estimated that 10% of the population consume twice the average quantity of a particular food class and 2.5% consume three times the average (Sherlock & Walters, 1983). Estimates of the daily cadmium intake in areas of Japan considered normal are consistently higher than in other parts of the world and generally range from 30 to 50 μg. In areas of elevated exposure, average daily intakes range from 150 to 250 μg (Tables 8 and 9).

Of particular interest is the village of Shipham, United Kingdom, where markedly elevated soil cadmium levels are present. Cadmium levels in locally grown vegetables have also been found to be elevated and ranged from 5 to 20 times above

Table 8. Estimates of average daily dietary intake of cadmium based on food analysis in various countries

| Country | Method of sampling[a] | Estimates (μg cadmium per day) | Reference |
|---|---|---|---|
| ***Areas of normal exposure*** | | | |
| Belgium | D | 15 | Buchet et al. (1983) |
| Finland | M | 13 | Koivistoinen (1980) |
| Japan | D | 31 | Yamagata & Iwashima (1975) |
| Japan | D | 48 | Suzuki & Lu (1976) |
| Japan | D | 49 | Ushio & Doguchi (1977) |
| Japan | D | 35 | Iwao (1977) |
| Japan | M | 49 | Ohmomo & Sumiya (1981) |
| Japan (mean of 3 areas) | D | 59 | Iwao et al. (1981a) |
| Japan | D | 43.9 (males)<br>37.0 (females) | Watanabe et al. (1985) |
| New Zealand | D | 21 | Guthrie & Robinson (1977) |
| Sweden | D | 10 | Wester (1974) |
| Sweden | M | 17 | Kjellström (1977) |
| United Kingdom | M, D | 10-20 | Walters & Sherlock (1981) |
| USA | M | 41 | Mahaffey et al. (1975) |
| ***Areas of elevated exposure*** | | | |
| Japan | M | 211-245 | Japan Public Health |
| Japan | D | 180-391 | Association (1970) |
| Japan (mean of 3 areas) | D | 136 | Iwao et al. (1981a) |
| United Kingdom | M | 36 | Sherlock et al. (1983) |
| United Kingdom | D | 29 | Sherlock et al. (1983) |
| USA | D | 33 | Spencer et al. (1979) |

[a] M - Sample of foodstuffs individually analysed; market basket method
D - Duplicate portion study

normal values (Table 6). Three dietary and crop sampling surveys were performed to estimate heavy metal intake from both fresh and cooked food (Sherlock et al., 1983). A duplicate portion study

Table 9. Estimates of average daily faecal cadmium elimination in various countries

| Country | Subjects investigated | Estimates (μg cadmium/day) | Reference |
|---|---|---|---|
| ***Areas of normal exposure*** | | | |
| Federal Republic of Germany | 23, sex and age of subjects not given | 31 | Essing et al. (1969) |
| Japan | 12 men, 50-59 years | 81 | Haga & Yamawaki (1974) |
| Japan | 13 women, 50-59 years | 56 | Haga & Yamawaki (1974) |
| Japan | 2 men, 35 and 37 years (60 specimens) | 36 | Suzuki & Lu (1976) |
| Japan | 7 men, 21-22 years (35 specimens) | 41-79 | Tati et al. (1976) |
| Japan | 64 men and women, 50-69 years | 41 | Kojima et al. (1975) |
| Japan | | 24-36 | Iwao (1977) |
| Japan (rural area) | 30 men, 50 years and over | 49 | Tsuchiya & Iwao (1978) |
| Sweden | 4 adults (2 men, 23 years; 2 women, 28 and 31 years) | 6-13 | Wester (1974) |
| Sweden | 70 men and 10 women (3-day collect) | 18 | Kjellström et al. (1978) |
| USA | 216 (men and women) | 10-15 | Kowal et al. (1979) |
| ***Areas of elevated exposure*** | | | |
| Japan, Kosaka | 40 men, 50-69 years | 149 | Haga & Yamawaki (1974) |
| Japan, Kosaka | 47 women, 50-69 years | 177 | Haga & Yamawaki (1974) |
| Japan, Kosaka | 118 men and women | 146 | Kojima et al. (1977) |
| Japan, Kakehashi, Kosaka, Tsushima (rural areas) | 30 men, 50 years and over | 149 | Iwao et al. (1981b) |
| New Zealand, Bluff | 45 men and women, 20-70 years | 50-500 | McKenzie et al. (1982) |

and two market basket studies were performed to coincide with periods of significant consumption of home-grown vegetables. The total cadmium dietary intake estimated from the market basket studies averaged 36 μg per day, of which 14 μg per day was contributed by locally grown fruit and vegetables. The duplicate portion study gave an average total cadmium intake of 29 μg per day, of which 17 μg per day was attributed to locally grown fruit and vegetables. Four individuals from the study population showed cadmium intakes greater than 400 μg/week. Both methods indicated that cadmium intakes in Shipham were higher than the United Kingdom average. However, exposures were not as high as would have been expected, considering the extent of cadmium contamination of the local vegetables, suggesting that most inhabitants did not rely heavily on local crops.

## 5.3 Total intake and uptake of cadmium from all environmental pathways

### 5.3.1 *General population, uncontaminated areas*

Assuming an air cadmium concentration of 10 ng/m$^3$ for both indoor and outdoor air and a daily inhalation rate of 15 m$^3$ for an adult, the average intake of cadmium from the atmosphere would be 0.15 μg, of which about 25% (Friberg et al., 1974) or 0.04 μg will be absorbed. Smoking a pack of 20 cigarettes daily can result in the inhalation of 2-4 μg cadmium, the amount varying considerably according to the country or origin of the tobacco. Of this amount, 25-50% may be absorbed via the lungs, resulting in an uptake of 1-2 μg, a much larger amount than from air alone. Those individuals who smoke two or more packs of cigarettes daily will absorb correspondingly greater quantities of cadmium.

Cadmium intake from drinking-water based on a daily consumption of 2 litres is usually less than 1 μg. Average daily intake from food in most countries is probably at the lower end of the range of 10-25 μg. At an absorption rate of 5%, daily uptake from water and food would be 0.6-1.3 μg cadmium. Thus, heavy smokers from the general population in uncontaminated areas may absorb more cadmium from the inhalation pathway than from dietary sources.

### 5.3.2 *General population, contaminated areas*

Airborne cadmium in contaminated areas may reach levels of 0.5 μg/m$^3$, which would lead to a daily inhalation of 7.5 μg and an absorption of about 2 μg. For smokers, the contribution from tobacco at 1-2 μg for every pack will not be changed, leading to a total uptake of 3-4 μg from inhalation in such individuals.

The intake of cadmium from food and water varies considerably and is related to both the extent of contamination and the reliance on locally grown food items or local water supplies. Daily intakes of 150-200 μg have been reported in contaminated areas where the majority of the staple food items were of local origin. At an absorption rate of 5%, the daily uptake from diet would be 8-10 μg. The total daily cadmium uptake will depend on the nature of cadmium contamination, i.e. whether food, water, and air levels are elevated, but is unlikely to exceed 20 μg.

Cadmium intake in children via the ingestion of household dusts is unlikely to be important except in the most contaminated localities.

### 5.3.3 *Occupational exposure to cadmium*

Inhalation of workplace air is the dominant exposure pathway. With air concentrations of 10-50 $\mu g/m^3$ and the inhalation of 10 $m^3$ air during a work-shift, the daily cadmium intake would be 100-500 μg. An absorption rate of 25% would thus lead to daily uptakes of 25-125 μg. Dust particles cleared from the lungs may be swallowed and dust-contaminated food items can also make a significant contribution to the ingestion pathway. At an absorption rate of 5% this could lead to an additional uptake 10-15 μg cadmium to the total uptake.

Tobacco carried by workers can become contaminated and may contribute up to 10 times more cadmium to the daily uptake than under normal conditions (Piscator et al., 1976).

## 5.4 Conclusions

The major route of exposure to cadmium for the non-smoking general population is via food; the contribution from other pathways to total uptake is small. Tobacco is an important source of cadmium uptake in smokers. In contaminated areas, cadmium exposure via food may be up to several hundred μg/day. In exposed workers, lung absorption of cadmium following inhalation of workplace air is the major route of exposure. Increased uptake in workers can also occur as a consequence of contamination of food and tobacco.

# 6. KINETICS AND METABOLISM IN LABORATORY MAMMALS AND HUMANS

## 6.1 Uptake

### *6.1.1 Absorption by inhalation*

Three processes in the lungs, i.e. deposition, mucociliary clearance, and alveolar clearance, determine the absorption of inhaled particles (Task Group on Lung Dynamics, 1966). Uptake into epithelial cells, interstitium or the systemic circulation depends on physical and biochemical processes in the respiratory tract after deposition (e.g., mechanical clearance, solubilization, and transport). The retained or accumulated dose at the local or systemic target site resulting from the deposited dose may eventually lead to biological effects. Length of exposure is of major importance for chronic effects, particularly lung cancer. Therefore, chronic effects might be expected to correlate with retained or accumulated dose rather than deposited dose.

Extrapolation models of inhaled cadmium dosage from animal models to humans and from high exposures (experimental) to low (environmental) must incorporate the above variables. When extrapolating from one species to another, specific pulmonary retention must be taken into account. In both acute and chronic inhalation exposures, a dose-response relationship is best described with accumulated rather than deposited dose (Oberdorster, 1988).

The absorption of cadmium compounds may vary greatly. As discussed in section 5.1.2, the proportion of particles in industrial air that are respirable, i.e. up to 5 μm MMAD, may vary widely (Materne et al., 1975). These particles will be deposited in the alveoli (Task Group on Lung Dynamics, 1966).

There are some empirical data on the overall absorption of cadmium. In various acute and chronic animals experiments, 5 to 20% of inhaled cadmium has been found to be deposited in the lungs (Friberg et al., 1986). Actual absorption may vary between 50 and 100% of the amount deposited and may continue for weeks after the deposition of a single dose. The absorption of an aerosol of cadmium chloride is higher than that of cadmium oxide, and alveolar absorption is higher after intratracheal instillation than after inhalation of an aerosol (Friberg et al., 1985).

If the particles are deposited in the alveoli, then the majority will sooner or later be absorbed, regardless of solubility. Cadmium chloride passes the alveolar-blood barrier with ease, although inhaled cadmium sulfide has a greater tendency to be retained in the lungs, indicating slower absorption. Three weeks after exposure of Syrian hamsters to cadmium chloride aerosol, about 25-35% of the initial lung burden was present in the liver, kidneys, and skull. The lungs still contained 50% of the initial lung burden at this time (Henderson et al., 1979).

Data on the respiratory absorption of cadmium in humans comes largely from comparisons of smokers and non-smokers. On the basis of data on organ burdens of cadmium and smoking history, Elinder et al. (1976) calculated that about 50% of the cadmium inhaled via cigarette smoke could be absorbed.

### 6.1.2 Absorption via the intestinal tract

Factors affecting the absorption of ingested cadmium include animal species, type of compound, dose, frequency of administration, age of experimental animals, pregnancy and lactation, presence or absence of drugs, and interactions of cadmium with various nutrients (Nomiyama, 1978). A study in which cadmium chloride was given in drinking-water to rats over a period of 12 months showed retention in the kidney and liver of less than 1% of the total amount ingested (Decker et al., 1958). There have been many reports of single exposure studies. These may be summarized as follows: the individual absorption of cadmium nitrate or chloride after single exposure ranges from 0.5 to 8% (Friberg et al., 1974). Limited observations in humans given radioactive cadmium indicate that the average absorption is about 5% (Kitamura, 1972; Rahola et al., 1972; Yamagata et al., 1974; Flanagan et al., 1978).

Metallothionein-bound cadmium in food does not appear to be absorbed and/or distributed in the same way as inorganic cadmium compounds. Mice exposed to cadmium-thionein (Cherian et al., 1978) had lower blood and liver cadmium levels but a higher kidney level than mice exposed to the same amount of cadmium as the chloride. Similar results were reported by Sullivan et al. (1984) in mice fed inorganic or oyster-incorporated radiolabelled cadmium. Cadmium in New Zealand Bluff oysters is to a great extent bound to a metallothionein-like protein (Nordberg et al., 1986). However, in other species of oysters, most of the cadmium is bound to proteins with relative molecular masses above 50 000 and lesser amounts to small proteins (< 3000) (Casterline & Yip, 1975; Kodama et al., 1978). Bluff oyster fishermen with an extremely high cadmium intake (up to 500 μg per day) from oyster consumption were found to have increased blood and urine cadmium levels (Sharma et al., 1983), but the increase was not as great as expected from the total cadmium ingested. This indicates that in humans, as in other animal species, metallothionein-bound cadmium in food may be dealt with in a different way from other cadmium compounds.

There are no data from humans studies showing a relationship between gastrointestinal absorption of cadmium and age. Studies on mice reported by Matsusaka et al. (1972), however, show approximately 10% whole body retention 2 weeks after ingestion for young mice, while the corresponding figure for adult mice was 1%. Kello & Kostial (1977) and Engstrom & Nordberg (1979b) also demonstrated that neonatal mice absorbed cadmium to a much greater extent than adult mice.

Diets with low levels of calcium and protein promote increased absorption of cadmium through the intestinal tract, up to 3 times the absorption having been noted in several studies in experimental animals (Friberg et al., 1974, 1975). It has also been shown that iron-deficient animals may have a higher absorption of cadmium (Hamilton & Valberg, 1974), and these findings have been confirmed in humans (Flanagan et al., 1978). Women with low body iron stores, as reflected by low serum ferritin levels, had on average, a gastrointestinal absorption rate twice as high (about 10%) as a control group of women. The highest individual absorption rate was about 20%. Interrelationships between cadmium exposure and the absorption of copper, zinc, and calcium will be discussed in section 7.5.

### 6.1.3 Absorption via skin

Limited skin penetration (1.8% per 5 h) of soluble cadmium compounds can take place when they are applied as a solution to the skin (Skog & Wahlberg, 1964). The dermal absorption rate was estimated by Kimura & Otaki (1972) in shaved rabbits and nude mice painted with an aqueous solution of cadmium chloride. Rabbits painted 5 times in 3 weeks showed a combined cadmium accumulation of 0.4-0.6% of the amount applied, and mice painted 1-4 times in one week showed an accumulation of 0.2-0.8% of the applied dose.

### 6.1.4 Transplacental transfer

The movement of cadmium through the placenta is limited. It has been shown that cadmium given to pregnant mice and hamsters during early pregnancy reaches the yolk sac and the primitive gut of the embryo, which are connected by the vitelline duct (Dencker et al., 1983). However, after closure of the vitelline duct during the later stages of pregnancy, very little cadmium reaches the fetus (Ahokas & Dilts, 1979). Sonawane et al. (1975) found that less than 0.02% of the total dose of cadmium injected intravenously into rat dams reached the fetus.

The cadmium concentration of the human placenta is usually about 5-20 μg/kg wet weight (Thieme et al., 1977; Copius-Peereboom et al., 1979). The placentas of women who smoke during pregnancy have higher levels than those of non-smokers (Copius-Peereboom et al., 1979).

Fetal (umbilical cord) blood cadmium levels are about 40-50% less than those of maternal blood. However, levels of the metabolically related essential metals zinc and copper in fetal blood are similar to or higher than those in maternal blood, this resulting in a fetal-maternal gradient (Lauwerys et al., 1978; Roels et al., 1978; Kuhnert et al., 1982; Korpela et al., 1986). The effectiveness of the gradient or its mechanism, as well as the potential toxicity of cadmium to the fetus, is not really known. Transplacental transport of cadmium is minimized in the normal healthy placenta presumably by the binding of cadmium to metallothionein. Metallothionein also serves as a site for intracellular zinc and copper sequestration. These observations suggest that there is a selective barrier to transplacental transport of cadmium. This is not the case with lead or mercury where fetal blood levels are similar to maternal levels (Lauwerys et al., 1978; Korpela et al., 1986).

## 6.2 Transport

Human data on the transport of cadmium from the site of absorption to the various organs are not available. This section is, therefore, based on animal studies, although there are some indications that similar mechanisms operate in humans. For instance, metallothionein has been isolated from human tissues (section 6.8) and has been measured in human plasma (Nordberg et al., 1982), where it binds cadmium being transported between tissues.

A study on dogs showed that, immediately after parenteral administration, most of the cadmium was present in the plasma (Walsh & Burch, 1959). This has been verified in a large number of animal studies (Friberg et al., 1974). Plasma concentrations decrease rapidly during the first hours after injection, reaching a level that is less than 1% of the initial value at 24 h, and this level then decreases much more slowly. During the early, fast-elimination phase, cadmium in mouse plasma is mainly bound to plasma proteins with a molecular weight of 40 000 to 60 000 (probably albumin), whereas in the slower phase (more than 24 h after injection), it is partly bound to a low molecular weight (LMW) protein of the same size as metallothionein (Nordberg, 1978). After rats were repeatedly exposed by subcutaneous injection (up to 14 weeks), the cadmium in plasma was partly bound to proteins with a molecular weight of 40 000 to 60 000 and partly to a LMW protein with a molecular weight similar to that of metallothionein (Cherian & Shaikh, 1975; Shaikh & Hirayama, 1979). The proportion of plasma cadmium bound to metallothionein and larger proteins, respectively, is considered to vary with the length and type of exposure. It is likely that the LMW cadmium-binding protein is in fact metallothionein, since it was shown by Vander Mallie & Garvey (1979) by a radioimmunological technique that the metallothionein concentration increased in the plasma of rats given 40 intraperitoneal injections of cadmium chloride in saline (0.12 mg/day, five days/week).

The concentration of cadmium in blood cells increases rapidly after a single intravenous injection (1 mg/kg body weight) and, within a few hours, reaches a first peak concentration exceeding that of the plasma. Although the levels of cadmium per cell may be 10 times higher in leucocytes than in red cells, the total cadmium in the leucocyte portion of the blood is negligible compared to that in the red cells (Garty et al., 1981).

Cadmium in erythrocytes may partly be bound to haemoglobin (Carlson & Friberg, 1957; Nomiyama et al., 1978a). However, during the first hour after a single subcutaneous injection, a large proportion of the cadmium in erythrocytes is bound to proteins with a molecular weight larger than haemoglobin (Nordberg, 1972). Between 96 and 196 h after a single injection (1 mg/kg body weight), it has been shown in mice (Nordberg, 1972), as well as in rats (Garty et al., 1981), that cadmium is also bound to a LMW protein. Whether this protein is identical with metallothionein is uncertain (Nordberg, 1984). A part of the erythrocyte cadmium in rats was also found in erythrocyte ghosts (membranes)

(Garty et al., 1981). When mice were exposed by subcutaneous injection to cadmium chloride (0.25 mg/kg body weight) for periods of between 6 days and 5 months (Nordberg et al., 1971), most of the erythrocyte cadmium was bound to a LMW protein similar to metallothionein.

Since metallothionein-bound cadmium is quickly cleared from the plasma by the kidneys (Nordberg & Nordberg, 1975; Vostal, 1976), this LMW fraction may be of great importance for the transport of cadmium from liver to kidney during long-term exposure. Hepatic metallothionein may be released into the blood in the same manner as hepatic enzymes and transported to the kidney and urine in some types of hepatic disorders (Tanaka, 1982).

## 6.3 Distribution

### 6.3.1 *In animals*

The highest cadmium levels in exposed animals are generally found in the liver and renal cortex. However, the distribution in the body varies according to the route of administration.

#### 6.3.1.1 *Single exposure*

Studies on various species have shown that, after a single administration of cadmium by the oral or parenteral routes, the highest organ burden of cadmium is initially found in the liver. However, kidney levels of cadmium increase for up to 8 months after exposure and may then exceed the liver levels (Gunn & Gould, 1957). The pancreas and spleen also show relatively high concentrations (Nordberg & Nishiyama, 1972). This topic has been reviewed by Friberg et al. (1974) and Nomiyama (1978).

#### 6.3.1.2 *Repeated exposure*

The literature on the fate of cadmium in animals after repeated exposure via various routes has been reviewed by Friberg et al. (1985) and, with emphasis on Japanese studies, by Nomiyama (1978). Liver cadmium levels increase rapidly, and a re-distribution of cadmium to the kidney occurs over a period of time. The higher the intensity of exposure, the higher the initial liver-to-kidney concentration ratio. The route of administration has been shown to be an important variable affecting the distribution of cadmium. When cadmium was administered subcutaneously, 11 times more was deposited in the liver than in the kidneys, whereas orally administered cadmium was distributed almost equally between these two organs (Nomiyama et al., 1976).

When rabbits were injected subcutaneously with 0.5 mg cadmium chloride daily, concentrations of cadmium in the liver and renal cortex reach a peak after about 10 and 15 weeks exposure, respectively. In cases of renal damage, urinary excretion increases (section 6.5.1.1) and the renal and liver concentrations decrease (Bonnell et al., 1960; Nomiyama et al., 1982b).

### 6.3.2 In humans

Cadmium is stored to the greatest extent in the liver and kidneys, the renal cortex showing the highest concentration in people who have not been exposed to excess cadmium (Friberg et al., 1974). The lowest concentrations (wet weight) are found in the brain, bone, and fat (Sumino et al., 1975; Cherry, 1981). Cadmium levels in the organs of second and third trimester fetuses (Chaube et al., 1973) and in newborn babies and young children (Henke et al., 1970) are lower by three orders of magnitude than in adult females. The placenta contains somewhat higher concentrations than maternal blood, brain or fat (section 6.1.4).

It has been calculated that about a third of the body burden in a non-smoking male from the USA is in the kidney and about a quarter in the liver and muscles. These are the tissues with the longest biological half-time of cadmium (section 6.6.2). In spite of the low cadmium concentration in the muscles, the contribution to the total body burden is great due to the large weight of the muscles. Other tissues that contribute significantly to body burden are bone, skin, and fat (Kjellström (1979).

In cadmium workers and people in the general environment exposed to high levels of cadmium, the liver or kidneys show the highest concentration, depending on exposure time, exposure levels, and the level of renal function (Friberg et al., 1974, 1985).

## 6.4 Body burden and kidney burden in humans

The newborn baby is practically free of cadmium (section 6.1.4), and the concentrations of cadmium in the organs increase with age (Schroeder & Balassa, 1961; Anke & Schneider, 1974; Elinder et al., 1976; Tsuchiya et al., 1976; Kowal et al., 1979;

Chung et al., 1986). The accumulation in human liver and muscles is shown in Figs. 1 and 2, respectively. These data and those of Vahter (1982) (Fig. 3) reveal important differences between people from different countries. Great individual variation also exists, even among people from the same area (Tsuchiya et al., 1976). For example, the geometric mean concentration of cadmium in the renal cortex of 117 adults aged between 30 and 59 in Stockholm was 19 mg/kg (Elinder et al., 1976). The individual concentrations followed a log-normal distribution with a geometric standard deviation of 2.0. This means that about 15% of the population would have values higher than 38 mg/kg, and 2.5% values higher than 76 mg/kg. Similarly shaped distributions were found for the kidneys, liver, pancreas, and muscle (Tsuchiya & Iwao, 1978; Kowal et al., 1979; Vuori et al., 1979).

The critical organ in long-term exposure to low concentrations of cadmium is the kidney (section 6.7). Initial cadmium-induced effects occur mainly in the proximal tubules, situated in the cortex of the kidney. Therefore, cadmium concentrations in the renal cortex and the distribution of cadmium within the kidney are of key importance. The weight of the renal cortex is about 2-3 times greater than the weight of the renal medulla, and early estimates of renal cortex cadmium concentrations (Friberg et al., 1974) were 1.5 times higher than the whole kidney concentrations. A recent study specifically aimed at measuring this concentration ratio (Svartengren et al., 1986) yielded an average value of 1.25 for people aged 30-50. This figure will be used in this document when it is necessary to recalculate whole kidney concentrations from renal cortex concentrations for that age group. This is the best estimate available at present, although the ratio may vary depending upon the age groups and racial types studied.

Table 10 shows the average cadmium concentrations in renal cortex and liver for the 20-59-year age group, and includes the major studies that have reported age-specific data. Unfortunately, no information on smoking habits was given in most studies. It has been shown that smoking cigarettes may significantly increase the body burden of cadmium (Lewis et al., 1972). Elinder et al. (1976) showed that Swedish smokers have, on average, about twice the tissue cadmium concentration of non-smokers. Except for the data from India (Fig. 3), there appears to be a constant difference (10 mg/kg) between smokers and non-smokers in the cadmium concentrations of the renal cortex. Fig. 3 also shows that the 90th percentile is usually about twice the geometric mean value. In most countries referred to in Table 10, the average cortex

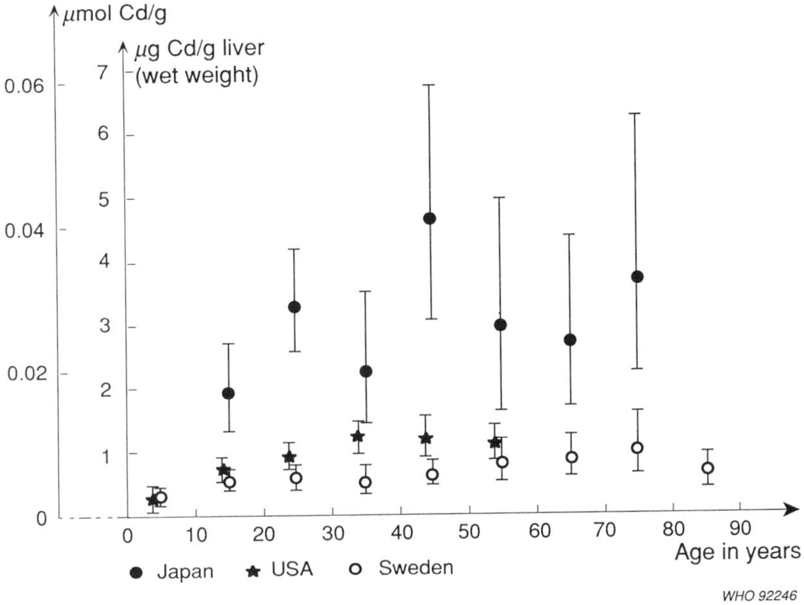


Fig. 1. **Average liver cadmium concentrations in men from Japan, USA, and Sweden (95% confidence intervals of mean indicated)**
From: Kjellström (1979)

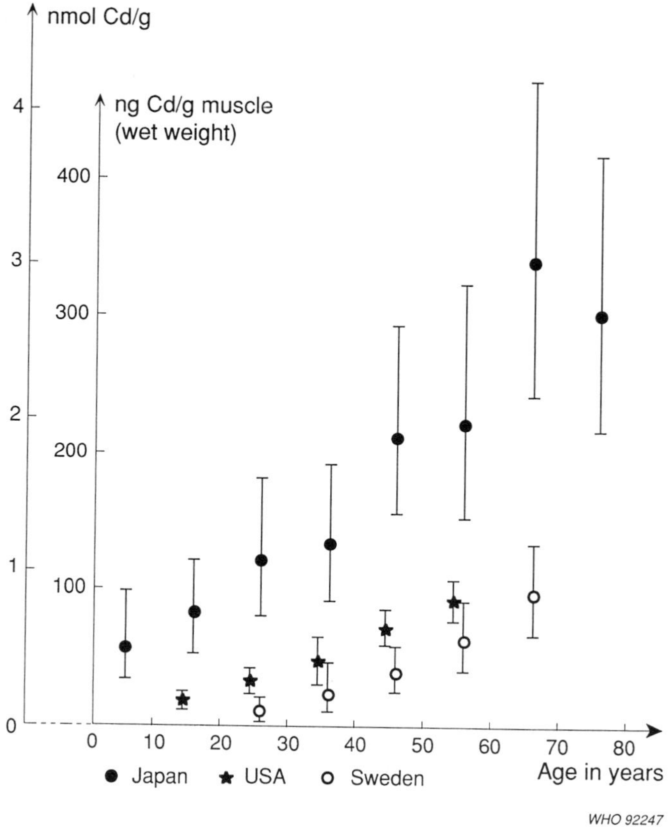


Fig. 2. Average muscle cadmium concentrations in men from Japan, USA, and Sweden (95% confidence intervals of mean indicated) From: Kjellström (1979)

cadmium concentration was in the range 10-40 mg/kg, while in Japan values of between 50 and 100 mg/kg were reported. In workers highly exposed to cadmium, but without functional impairment of the kidney, concentrations in the renal cortex may range from 180 to 450 mg/kg wet weight. In cases where there is severe renal dysfunction, the cadmium concentrations are generally lower and range between 20 and 120 mg/kg wet weight, i.e. they are of the same magnitude as those of the general population (Friberg et al., 1974). This seemingly paradoxical relationship is discussed in more detail in section 6.7.

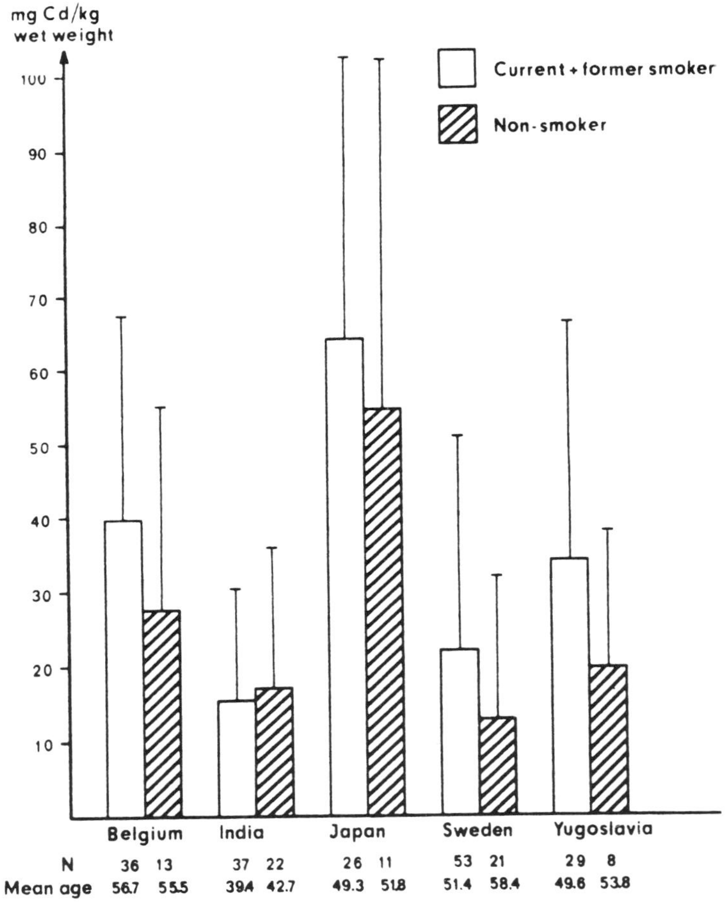


Fig. 3. Concentration of cadmium in kidney cortex (geometric mean values with 1.28 times the geometric standard deviations indicated) in relation to smoking habits among adults (30-69 years of age) in Belgium, India (data from Ahmedabad, Bangalore, and Calcutta pooled), Japan, and Yugoslavia. Swedish data are derived from Elinder et al. (1976). Numbers of smokers (including former smokers) and non-smokers as well as mean age in each subgroup are indicated under the bars. From: Vahter (1982)

If exposure to cadmium throughout life remains constant and low in amount, the concentrations in the kidneys become higher (by about 10-20 times) than those in the liver. Average liver cadmium concentrations are about 1-2 mg/kg wet weight at age 50 in some European countries and the USA, but in Japan average

Table 10. Cadmium concentrations in the renal cortex and liver of people from various geographical areas[a]

| Country | Number | Sex | Age group | Smoking category | Renal cortex cadmium level (mg/kg wet weight) | Liver cadmium level (mg/kg wet weight) | Reference |
|---|---|---|---|---|---|---|---|
| Belgium (Liege) | 51 | M, F | 40-59 | mixed | 46[b] | - | Vahter (1982) |
| German Democratic Republic | 20 | M | 40-59 | mixed | 22[c] | - | Anke & Schneider (1974) |
| | 20 | F | 40-59 | mixed | 11[c] | - | Anke & Schneider (1974) |
| India | 26 | M, F | 40-59 | mixed | 24[b] | - | Vahter (1982) |
| Israel (Jerusalem) | 11 | M, F | 40-59 | mixed | 28[b] | - | Vahter (1982) |
| Japan (Kobe) | 6 | M, F | 50-59 | mixed | 54 | 5.0 | Kitamura et al. (1970) |
| Japan (Kanazawa) | 9 | M, F | 40-59 | mixed | 95 | 10 | Ishizaki (1972) |
| Japan (Tokyo) | 17 | M, F | 40-59 | mixed | 99 | 5.7 | Tsuchiya et al. (1976) |
| Japan (Tokyo) | 23 | M, F | 40-59 | mixed | 76[b] | - | Vahter (1982) |
| Sweden (Stockholm) | 83 | M, F | 40-59 | mixed | 20 | 0.76 | Elinder et al. (1976) |
| | 45 | M | 40-59 | mixed | 18 | 0.69 | Elinder et al. (1976) |
| | 38 | F | 40-59 | mixed | 23 | 0.88 | Elinder et al. (1976) |
| | 7 | M, F | 40-59 | non-smoker | 11 | 0.51 | Elinder et al. (1976) |
| | 28 | M, F | 40-59 | smoker | 23 | 0.97 | Elinder et al. (1976) |

Table 10 (contd).

| Country | Number | Sex | Age group | Smoking category | Renal cortex cadmium level (mg/kg wet weight) | Liver cadmium level (mg/kg wet weight) | Reference |
|---|---|---|---|---|---|---|---|
| USA (North Carolina) | 19 | M | 40-59 | mixed | 27[d] | 2.2 | Hammer et al. (1973) |
| | 10 | F | 40-59 | mixed | 23[d] | 2.2 | Hammer et al. (1973) |
| | 10 | M | 40-79 | non-smoker | 14[d] | 1.6 | Hammer et al. (1973) |
| | 18 | M | 40-79 | smoker | 28[d] | 3.2 | Hammer et al. (1973) |
| USA (Dallas) | 58 | M | 40-59 | mixed | 29 | 1.4 | Kowal et al. (1979) |
| | 47 | M | 20-59 | non-smoker | 13 | 1.02 | Kowal et al. (1979) |
| | 115 | M | 20-59 | smoker | 24 | 1.31 | Kowal et al. (1979) |
| USA (Baltimore) | 10 | M, F | 40-59 | mixed | 30[b] | - | Vahter (1982) |
| Yugoslavia (Zagreb) | 28 | M, F | 40-59 | mixed | 38[b] | - | Vahter (1982) |

[a] The cadmium concentrations are arithmetic mean values and have been rounded off.
[b] Original data were geometric means. Adjusted according to the findings of Elinder et al. (1976) (x 1.18).
[c] Original data were for whole kidney (dry weight). Data adjusted for whole kidney (x 1.25) and dry weight (x 0.21).
[d] Original data were based on ash weight. Data adjusted for ash weight (x 0.011).

concentrations are between 5 and 10 mg/kg (Table 10). Although renal concentrations generally decrease after age 60 (Fig. 4), liver concentrations reach a plateau but do not show a clear decrease in aged populations (Fig. 1). In exposed workers, liver concentrations from 20 to about 300 mg/kg have been recorded and in Itai-itai patients they are between 63 and 132 mg/kg (Friberg et al., 1974). In people with severe cadmium-induced renal dysfunction, kidney cadmium levels are low, but the liver levels may be very high (Ishizaki, 1972).

*In vivo* neutron activation analysis (see section 2.2.3.3) has recently been used to measure kidney and liver cadmium concentrations in exposed workers. In one study, the detection limits were about 15 mg/kg for kidney and 1.5 mg/kg for liver (Ellis et al., 1981a), while in another study detection limits were higher by a factor of 2 for kidney and 5 for liver (Roels et al., 1981b). Thus, this method is still not sufficiently sensitive to measure *in vivo* tissue levels in a "normal" population. Liver levels of up to 120 mg/kg and kidney cortex levels of up to 600 mg/kg have been found among cadmium workers (Ellis et al., 1981a). A decreasing trend of the kidney levels after a maximum at about 300 mg/kg in kidney cortex was evident. At this point, the liver level was about 30 mg/kg (Fig. 5). A very similar situation was found in another factory (Roels et al., 1981b); most workers with high liver cadmium levels had low kidney levels, and then also showed elevated urinary excretion of $\beta_2$-microglobulin. In exposed workers, the average ratio of the cadmium concentration in the renal cortex to that in the liver has been reported to be about 8 (Ellis et al., 1981a) or 7 (Roels et al., 1981b), values that are lower than for the general population (Table 9). This corresponds to animal data (section 6.3.1.2) showing a greater proportion of accumulated cadmium in the liver when the exposure level increases.

The total body burden of cadmium in a middle-aged person within the general population is about twice the amount in kidneys and liver together (Table 10), i.e. 5-7 mg in a non-smoker in Europe or the USA and 8-13 mg in a smoker (Kjellström, 1979). In Japan, higher body burdens have been reported (Tipton et al., 1960; Ishizaki et al., 1971; Sumino et al., 1975; Tsuchiya et al., 1976). An extensive review of data from several countries (Cherry, 1981) found total body burdens to lie within the range 5-20 mg.

In conclusion, the average total body burden of a person of 50 years of age, living in an area not subject to pollution, varies

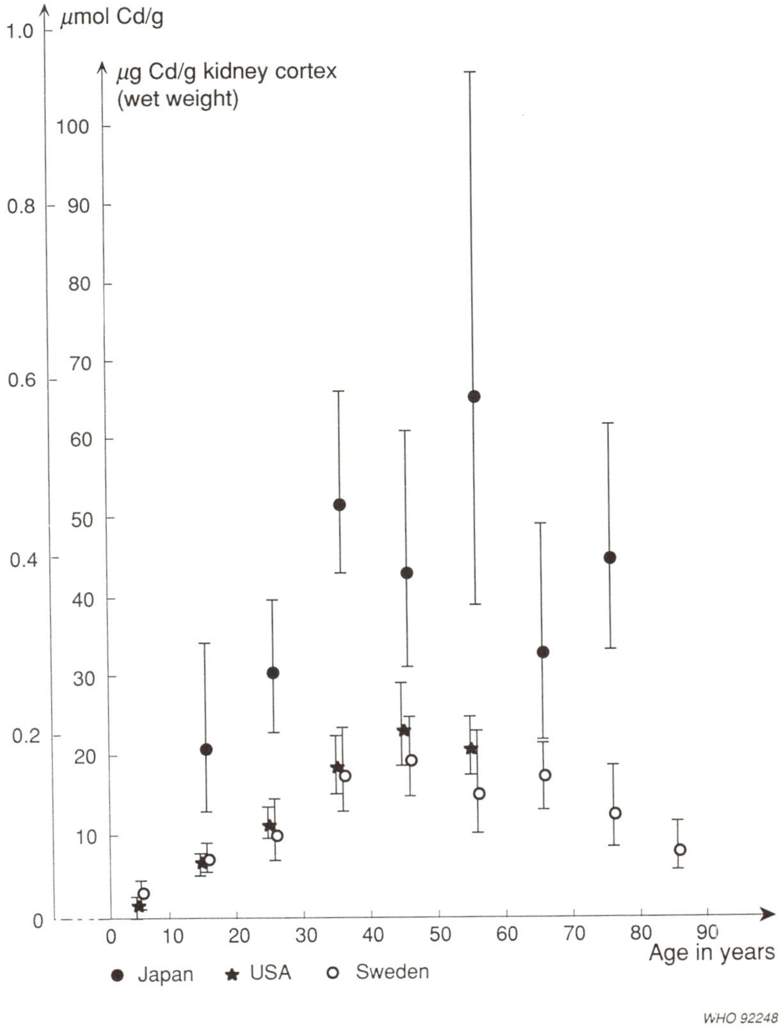


Fig. 4. Average cadmium concentrations in the renal cortex of men from Japan, USA, and Sweden (95% confidence intervals of mean indicated From: Kjellström (1979)

within the range 5-20 mg in different regions of the world, and the average cadmium concentration in the renal cortex varies within the range 11-100 mg/kg wet weight. There is a great individual variation, and the 90th percentile in those groups studied is about twice the median value.

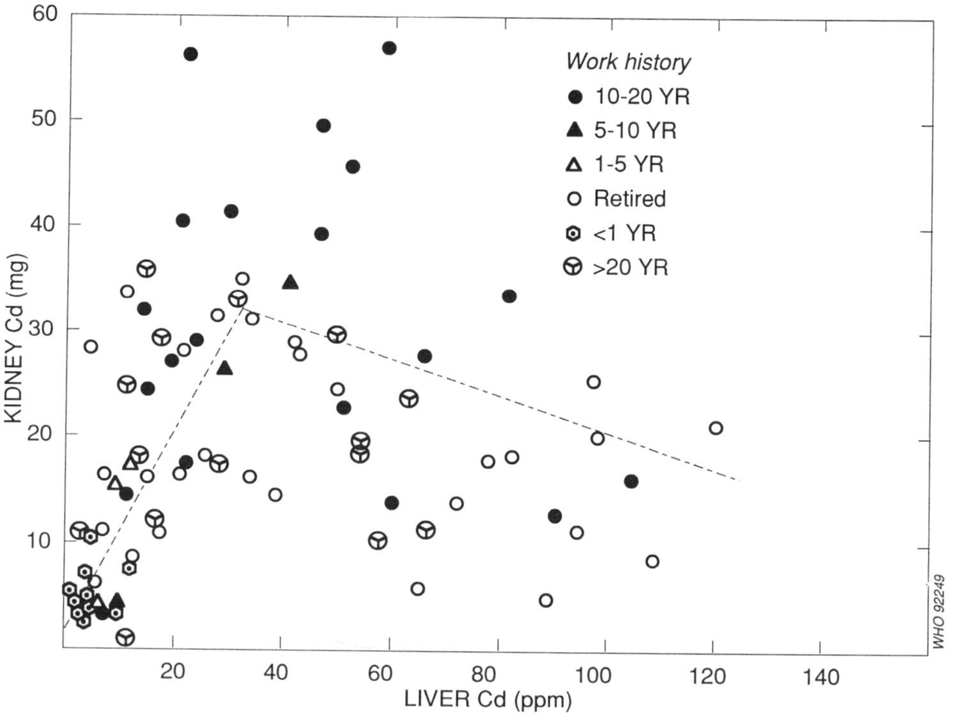


Fig. 5. Relation between kidney and liver cadmium levels for the general population (the dashed line represents the two components of a linear model fit to the data)

Smoking increases the body burden. After long-term low-level exposure, about half the body burden of cadmium is localized in the kidneys and liver, a third of the total being in the kidneys. At higher levels of exposure, a greater proportion of the body burden is found in the liver. After the development of severe cadmium-induced renal dysfunction, cadmium is lost from the renal tissue.

## 6.5 Elimination and excretion

### 6.5.1 Urinary excretion

#### 6.5.1.1 In animals

Nordberg (1972) demonstrated that after subcutaneous injection for up to 24-25 weeks, the average daily urinary cadmium excretion (on a group basis) in mice, prior to the onset of tubular proteinuria, represented about 0.01-0.02% of the body burden (section 6.7.1). Elinder & Pannone (1979) showed that one month after repeated subcutaneous exposure ceased, the excretion was only 0.001% of the body burden.

Similar low excretion rates have been found in rabbits given subcutaneous injections (Nomiyama, 1973a; Nomiyama & Nomiyama, 1976a), and in rabbits (Nomiyama & Nomiyama, 1976a,b) and monkeys (Nomiyama et al., 1979, 1982a) given cadmium orally. In addition, it has been reported that, over a range of doses, an increase in urinary excretion of cadmium is associated with an increase of cadmium in the renal cortex (Nomiyama & Nomiyama, 1976a; Suzuki, 1980; Bernard et al., 1981).

Studies on several mammalian species, mainly involving repeated subcutaneous injection of cadmium salts, have shown that urinary excretion of cadmium increases slowly for a considerable time but, as kidney dysfunction develops, a sharp increase in excretion occurs in rabbits (Friberg, 1952; Axelsson & Piscator, 1966a; Nomiyama & Nomiyama, 1976a), mice (Nordberg & Piscator, 1972), and rats (Suzuki, 1980). This leads to a decrease in renal and liver cadmium concentrations (Axelsson & Piscator, 1966a; Suzuki, 1980; Nomiyama & Nomiyama, 1976a).

When renal tubular lesions were induced by uranyl acetate injections in animals previously exposed to cadmium, there was no increase in urinary cadmium excretion (Nomiyama & Nomiyama, 1976a) or decrease in the level of cadmium in the renal cortex.

This contrasts with the increase in cadmium excretion brought about by cadmium-induced tubular lesions.

#### 6.5.1.2 In humans

Several studies have shown that in the general population urinary cadmium excretion increases with age (Katagiri et al., 1971; Tsuchiya et al., 1976; Elinder et al., 1978; Kowal et al., 1979) (Fig. 6), this increase coinciding with the increased body burden. Smokers have higher urinary excretion than non-smokers (Elinder et al., 1978; Kowal et al., 1979). The mean concentration of urinary cadmium in such groups of people not exposed to high cadmium levels is < 0.5-2.0 μg/litre or approximately 0.01% of the total body burden.

Increased urinary cadmium excretion occurs when tubular proteinuria develops (Lauwerys et al., 1974a; Kojima et al., 1977). In cadmium exposed workers, high urinary cadmium concentrations in the absence of proteinuria can be found after only short exposures (Lauwerys et al., 1976, 1979a,b) (section 6.7.1).

Most of the cadmium in urine is probably transported bound to metallothionein. The urinary metallothionein concentration can now be measured quantitatively with a sensitive radioimmunoassay (Vander Mallie & Garvey, 1979). Using this technique, Tohyama et al. (1981b) found good correlation between urinary metallothionein and cadmium in 67 people exposed in the general environment, and Roels et al. (1983b) confirmed this correlation in 94 cadmium workers.

### 6.5.2 Gastrointestinal and other routes of excretion

It is extremely difficult to study gastrointestinal excretion after oral exposure, since it is not possible to distinguish net gastrointestinal excretion from unabsorbed cadmium in faeces.

Animal studies of gastrointestinal excretion following injections of cadmium (summarized by Friberg et al., 1974) generally show that a few percent of the dose is excreted in the faeces within the first few days after injection. The faecal excretion is initially higher than the urinary excretion after either single or repeated exposure (Nomiyama, 1978). The mechanism for such excretion probably involves a transfer of cadmium via the intestinal mucosa, but biliary excretion may also be involved. The

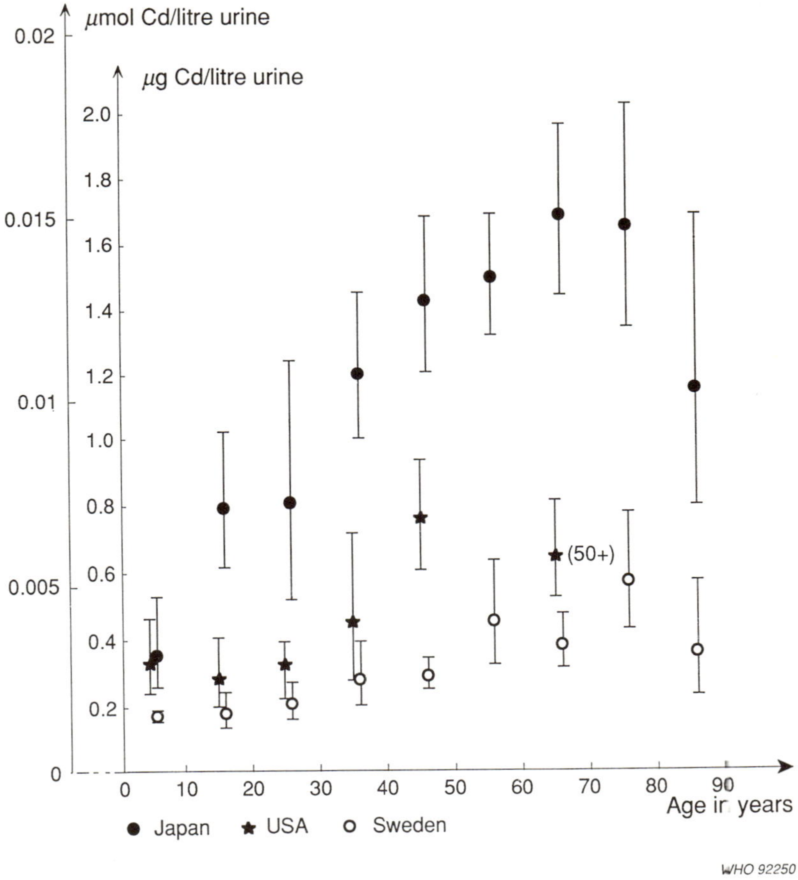


Fig. 6. Average urine cadmium concentrations in men from Japan and USA (smokers and non-smokers) and from Sweden (non-smokers only) (95% confidence intervals of mean indicated)

biliary excretion in the first 24 h after intravenous injection of cadmium is dependent on the dose (Cirkt & Tichy, 1974; Nomiyama, 1974; Klaassen & Kotsonis, 1977). In rats given 67, 90 or 120 µg cadmium (Cikrt & Tichy, 1974), the cumulative 24 h excretion reached 0.83% at the lowest dose and 5.68% at the highest dose. The highest excretion rate was detected between 15 and 30 min after dosing. It has been reported that after the initially rapid excretion the biliary excretion is 0.015-0.04% of the body burden per hour over three consecutive days (Nordberg et al., 1977; Elinder & Pannone, 1979). Biliary cadmium has been partially characterized as a glutathione complex (Cherian & Vostal 1977).

Both during and after parenteral exposure to cadmium, the total gastrointestinal cadmium excretion is considerably higher than the urinary excretion (Nordberg, 1972; Elinder & Pannone, 1979). A large proportion of the gastrointestinal excretion is directly related to the daily dose. After chronic exposure of rats, faecal excretion amounted to about 0.03% of the body burden, which was considerably more than the urinary excretion (Elinder & Pannone, 1979).

There are no available quantitative human data to indicate the net gastrointestinal excretion.

Cadmium is also eliminated through hair (Anke et al., 1976) and breast milk (Schroeder & Balassa, 1961), but these routes are of limited importance for total excretion and do not significantly alter the biological half-time.

## 6.6 Biological half-time and metabolic models

### 6.6.1 *In animals*

Several studies have been carried out in order to assess the biological half-times of cadmium in experimental animals. Various animals species, including mice, rats, rabbits, dogs, and monkeys, have been studied, and single exposures have normally been used. The reported half-times have varied from weeks to two years (or as long as half the life-span of the animal). The development of metabolic models has shown that the body contains several compartments for cadmium accumulation, each with a different half-time. Thus, in whole body half-time measurements, one may find several different half-times. In order to observe the slowest half-time components, it is necessary to study the animals for many months.

The biological half-time of cadmium in the kidney and whole body decreases when renal tubular dysfunction occurs because of increased urinary excretion (section 6.5.1). However, some studies have indicated that the biological half-time may change with dose and body burden of cadmium even before renal damage occurs. For instance, Engstrom & Nordberg (1979a) reported that in mice half-time increased with increasing single or repeated oral dose.

In these studies, body burden and renal burden were considerably lower than the maximum that can be reached in long-term exposure. Nomiyama (1978) reported that half-time decreased with increasing dose when animals with the shortened half-time had reached the maximum renal burden. Even shorter half-times were reported when renal tubular dysfunction occurred after exposure to high doses (Nomiyama, 1978).

A number of studies on biological half-time and metabolic models for animals have been reviewed by Friberg et al. (1974) and Nomiyama (1978).

The wide difference in the results obtained by investigators may be explained by variations in exposure level and type, the different animal species used, and interactions between cadmium and other exposure factors. Reported half-times range from several weeks in mice to 22 years in monkeys (Friberg et al., 1974; Nomiyama et al., 1979; Nomiyama et al., 1984). The variations in half-times in specific tissues between different species or individuals may be due to variations in the production of metallothionein, which binds tissue cadmium and contributes to its retention.

### 6.6.2 In humans

Experimental and epidemiological evidence indicates strongly that the biological half-time in the whole body is extremely long (many years). Experimental evidence from one study (Shaikh & Smith, 1980), in which one subject was given radioactive cadmium and examined periodically for the next 2 years, showed a biological half-time of 26 years. In three similar studies, in which a small number of subjects were followed up for a limited period (about 100 days), half-times of 93-202 days were reported (Rahola et al., 1972; Flanagan et al., 1978; McLellan et al., 1978). Only one of these studies (Rahola et al., 1972) gave confidence limits for the estimated biological half-time (130 days to infinity).

Another approach to estimate the half-time used involves comparing total daily excretion with total body burden, applying a one-compartment model to the body as a whole (Friberg et al., 1974; Task Group on Metal Toxicity, 1976). A further approach analyses the accumulation in the kidney using a one-compartment model taking into consideration age-related variations in daily cadmium exposure and kidney weight (Tsuchiya & Sugita, 1971; Kjellström, 1971). More recently, an elaborate model has been developed that includes separate compartments for, for instance, kidney, liver, and blood, and incorporates age-related variations in daily intake, tissue weights, and renal function (Kjellström & Nordberg, 1978).

These models rely on many assumptions concerning the cadmium concentration in food, calorific intake, absorption rates, and other factors. Inevitably, the data produced by these models are only tentative, but they are important for future research. Using data from Japan on the accumulation of cadmium with age, Tsuchiya et al. (1976) used a series of one-compartment mathematical models developed by Tsuchiya & Sugita (1971) to estimate biological half-times of cadmium in various organs. These authors estimated the biological half-time in the kidneys to be about 17 years and that in the liver 7 years.

Elinder et al. (1976) used autopsy data from non-smokers in Sweden and a one-compartment model (Kjellström, 1971) to estimate the biological half-time in the renal cortex. They assumed that the daily intake of cadmium had doubled in 50 years (Kjellström et al., 1975a) and estimated the half-time to be 20-50 years (30 years being the best estimate).

Using an 8-compartment model (Kjellström & Nordberg, 1978), the biological half-times of cadmium in the liver and kidney were estimated to be 7.5 and 12 years, respectively (Kjellström & Nordberg, 1978). The longest half-time was calculated for the "other tissues" compartment. This included muscle tissue, which was found to have the longest half-time in an autopsy study (Kjellström, 1977). However, it should be pointed out that when using this type of model to simulate the chemobiokinetics of cadmium, the individual half-times of different tissues are less important than the dynamics of the model as a whole.

After high cadmium exposure, as occurs among certain industrial workers, the biological half-time may not be the same as that during normal exposure in the general environment. Current models, however, do not consider this factor. If the exposure level is very high, the ratio of rapid components to slow components may be altered. An example of this is the very high urinary excretion found after only a short exposure to high air cadmium levels (Lauwerys et al., 1976, 1979b). The biological half-time is also shorter if there is renal tubular dysfunction. Fletcher et al. (1982) carried out *in vivo* neutron activation analysis of the liver of 13 cadmium workers twice within a period of 3 to 4 years (the occupational cadmium exposure of these workers had ceased before the first analysis). Three workers showed proteinuria and an average cadmium half-time in the liver of 2 years. The other 10 workers had an average half-time of 6.4 years, nine of whom had an average half-time of 13.5 years and no proteinuria.

Jarup et al. (1983) studied the half-time of cadmium in blood of five smelter workers who had previously experienced high cadmium exposure. Repeated blood analysis carried out over a 10- to 13-year period revealed short-term (75-128 days) and a long-term (7.4-16 years) half-time components. The long-term component in two workers with proteinuria was shorter than in the other workers.

## 6.7 Biological indices of cadmium exposure, body burden, and concentrations in kidneys

There is no easy way to measure directly the whole body burden or concentrations of cadmium in different tissues of a living person. *In vivo* neutron activation methods have been used in special circumstances (Ellis et al., 1981a; Roels et al., 1981b; Tohyama et al., 1981a).

At present, it is necessary to study concentrations in easily available indicator media in order to evaluate exposure and accumulation of cadmium. The suitability of certain indicator media for such purposes is supported by studies on both animals and humans; urine, blood, faeces, and hair have all been used as indicator media. Methods for the biological monitoring of cadmium levels in blood and urine have been reviewed by Nordberg & Nordberg (1988) and WHO (1980).

### 6.7.1 Urine

The human and animal studies summarized in section 6.5.1 allow the following interpretation of the significance of cadmium in urine (Lauwerys et al., 1980b). In the absence of episodes of high-level exposure to cadmium and provided that cadmium-binding sites in the organism are not saturated and cadmium-induced nephropathy has not yet occurred, the urine cadmium level increases in proportion to the amount of cadmium stored in the body. In such situations, which prevail mainly in the general population and in workers moderately exposed to cadmium, there is significant correlation between urinary cadmium and cadmium in kidney. Episodes of high exposure to cadmium, however, may lead to a transient increased urinary excretion.

If exposure to cadmium has been excessive, the cadmium-binding sites in the organism become progressively saturated and, despite continuous exposure, the cadmium concentration in the renal cortex tends to plateau. Once this point is reached, the cadmium that is still absorbed cannot be further retained in the kidney and is rapidly excreted in the urine. Under these conditions, urinary cadmium is also influenced by the recent intake. The relative influence of the body burden and the recent exposure on urinary cadmium depends on the exposure intensity. If exposure continues, a certain percentage of individuals may develop renal damage. This is associated with a progressive loss of cadmium accumulated in the kidney, which gives rise to a further increase in urinary cadmium. Eventually, the amount of cadmium that can be released from the kidney decreases progressively and the urinary cadmium concentration follows the same trend. The changes in the urinary metallothionein level parallel those of cadmium (section 6.8.2).

In summary, several factors (duration and intensity of exposure to cadmium, the presence of renal dysfunction and its duration) must be taken into consideration when interpreting urinary cadmium (and metallothionein) levels.

### 6.7.2 *Blood*

Plasma cadmium levels are considered to be related to recent exposure but are often so low that they cannot be measured routinely (section 6.2). Most of the cadmium in the blood is in the erythrocytes.

Cadmium levels in whole blood mainly reflect the exposure during recent weeks or months. In cadmium workers, the level increases markedly within the first few months after occupational exposure starts (Kjellström & Nordberg, 1978; Lauwerys et al., 1979b). It is probable that a portion of the blood cadmium level reflects body burden rather than present exposure in view of the known transport of cadmium via blood from the liver to the kidneys and other tissues (section 6.2) and the long-term half-time component demonstrated by Jarup et al. (1983). Workers with relatively long exposure durations but whose cadmium exposure has ceased have elevated blood cadmium levels for several years (Friberg et al., 1974; Jarup et al., 1983).

Reports of blood cadmium levels in the general population have, in the past, often been unreliable, owing largely to the difficulties encountered in the analysis of blood cadmium (Lauwerys et al., 1975). However, improvements in analytical techniques have since been achieved through biological standards for blood and systematic quality assurance programmes (Stoeppler et al., 1979; Vahter, 1982). Average blood cadmium values up to 10 μg/litre or more have been reported in the past, but the analytical procedures used mean that the accuracy of these data is in doubt (Vahter, 1982; Friberg & Vahter, 1983).

Various aspects of blood cadmium analysis have been discussed in a UNEP/WHO study, which also included a quality assurance programme and data from 10 countries (Vahter, 1982; Friberg & Vahter, 1983). It was found that even in the countries with the highest blood cadmium levels, the average was less than 4 μg/litre. Furthermore, smokers were found to have higher values than non-smokers, and non-smokers, in most countries, had mean levels below 1 μg/litre. Although only slightly above 1 μg/litre, the levels for non-smokers in Japan were about twice as high as the levels in the USA, which probably reflects the difference in average daily cadmium intake via food (section 5.2.2).

Reported cadmium concentrations in the blood of exposed workers are generally between 5 and 50 μg/litre, but levels of between 100 and 300 μg/litre have resulted from extreme exposures (Roels et al., 1982; Hassler et al., 1983).

### 6.7.3 Faeces

Gastrointestinal absorption amounts to only a few percent of the cadmium ingested daily (section 6.1.2), and the quality of cadmium excreted gastrointestinally is small compared to the unabsorbed portion of ingested cadmium. Thus, daily faecal cadmium can serve as a good indicator of the daily amount of cadmium ingested via food and water or cleared from the lungs after occupational exposure to large dust particles. Faecal cadmium correlates very closely with daily energy intake (section 5.2.4), and average cadmium intake estimates for different countries agree well with reported average faecal cadmium amounts.

A portion of the cadmium in human faeces is related to the body burden (section 6.5.2). In workers with high body burdens but low daily cadmium intakes via food, this portion might be greater than the unabsorbed part of ingested cadmium, because the faecal excretion is similar to the urinary excretion.

### 6.7.4 Hair

Cadmium in hair is not a reliable indicator of either recent exposure or body burden. The main problem is external contamination of the hair, which cannot be distinguished from the endogenous cadmium (Nishiyama & Nordberg, 1972). However, a correlation was found between air cadmium levels of cities in the USA and cadmium levels in the hair of 10-year-old children living in these cities (Hammer et al., 1971). In a study of cadmium workers (Ellis et al., 1981b), a higher average hair cadmium level was found among exposed workers than among controls, but there was a poor relationship between hair cadmium and cadmium in the blood, urine, liver or kidney. Cadmium concentrations in the hair of people without excessive exposure are usually between 0.5 and 2 mg/kg.

## 6.8 Metallothionein

### 6.8.1 Nature and production

Metallothionein is a metal-binding protein of low molecular weight, which has a key role in the metabolism of cadmium. It is rich in cysteine but contains no aromatic amino acids or histidine (Kagi & Vallee, 1960, 1961).

This protein was identified for the first time by Margoshes & Vallee (1957) in horse kidney cortex. Its molecular weight is about 6600 (6000 for the apoprotein moiety, thionein), and it has a non-globular shape. On gel filtration, however, it moves like a spherical protein with a molecular weight of about 10 000 (Kagi & Nordberg, 1979). There have been several reports dealing with the function and biochemistry of metallothionein (Kagi & Nordberg, 1979; Brady, 1982; Foulkes, 1982; Webb & Cain, 1982; Kagi & Kojima, 1987).

Piscator (1964) suggested that metallothionein played a role in cadmium transport and detoxication, and it has subsequently been identified in human kidney and liver (Pulido et al., 1966; Chung et al., 1986) as well as in those of various experimental animals (Kagi & Nordberg, 1979).

The structure and genetic expression of mouse and human metallothionein have now been identified. Two major forms of metallothionein are present in most mammalian tissues, particularly liver and kidney, i.e. metallothionein I (Mt-I) and metallothionein II (Mt-II). Induction of synthesis is under the control of a large group of genes and is stimulated by glucocorticoids and the essential metals zinc and copper, as well as by the toxic metals cadmium and mercury (Karin et al., 1981; Karin & Richards, 1982). *In vitro* binding affinities have been demonstrated for a number of other toxic metals, including bismuth, cobalt, silver, and gold (Cherian & Nordberg, 1983). Metallothionein binds seven metal ions per molecule between two separate metal-cysteine clusters, and a single molecule may contain more than one metal, e.g., cadmium and zinc, mercury and copper.

### 6.8.2 The role of metallothionein in transport, metabolism, and toxicity of cadmium

Piscator (1964) suggested that some of the cadmium-binding metallothionein in the liver may migrate into the blood stream. As discussed in section 6.2, part of the plasma cadmium in animals exposed for a long time to cadmium is bound to a protein with the same molecular weight as metallothionein. When metallothionein-bound cadmium is present in the plasma, it is quickly cleared by glomerular filtration and reabsorbed in the renal tubules or excreted in the urine (Cherian & Shaikh, 1975; Nordberg et al., 1975; Webb & Etienne, 1977; Fowler & Nordberg, 1978).

At low levels of cadmium-metallothionein in the plasma, tubular reabsorption is almost complete, whereas the uptake in the tubular cells from the tubular fluid is saturated in the presence of high concentrations (Nomiyama & Foulkes, 1977; Foulkes, 1982). Thus, high urinary excretion of cadmium-metallothionein occurs shortly after the administration of larger doses, i.e. doses exceeding about 0.1 mg cadmium/kg body weight (Cherian & Shaikh, 1975; Nordberg & Nordberg, 1975).

It has been demonstrated in animal and *in vitro* tissue studies that metallothionein provides a protective role for cadmium toxicity (Cherian & Nordberg, 1983). Mice pretreated with cadmium have increased tolerance to subsequent cadmium exposure (Nordberg et al., 1971), and exposure to cadmium may protect from subsequent mercury toxicity (Piotrowski et al., 1974). In addition, the inhibition of certain mixed-function oxidases by cadmium is reduced by prior induction of metallothionein by cadmium in immature mice (Asokan et al., 1984). Pre-exposure of cultured kidney cells to cadmium protects from subsequent exposure (Cherian, 1980; Jin et al., 1986).

Nordberg et al. (1975) showed that metallothionein isolated from rabbit and mouse liver produced acute renal tubular cell toxicity when injected subcutaneously into mice. Further study (Cherian et al., 1976; Fowler & Nordberg, 1978; Squibb et al., 1982, 1984) suggested that parenterally administered cadmium-metallothionein enters proximal renal tubular lining cells in pinocytotic vesicles that fuse with lysosomes. The metallothionein is degraded, releasing cadmium into the cytosol and producing cellular degeneration and necrosis within 8-24 h. The renal tubular cell toxicity produced by metallothionein with different ratios of cadmium and zinc is proportional to the cadmium content of the metallothionein (Suzuki, 1982). Zinc-thionein does not have a similar effect.

The pathogenesis of renal tubular cell toxicity is thought to be related to non-metallothionein-bound cadmium, which becomes rapidly bound to existing metallothionein sites or induces the synthesis of new metallothionein (section 7.2.1.4).

The prevalence of nephrotoxicity rather than hepatotoxicity in chronic cadmium exposure may be due to several factors. Firstly, the release of hepatic cadmium-metallothionein or its presence in the blood can result in preferential accumulation of cadmium in the kidneys. Secondly, it has been shown in experimental animals that the kidney can accumulate metallothionein mRNA in response to cadmium exposure to only about half the level of the liver (Koropatnick & Cherian, 1988). Thus, the kidney may not be able to synthesize metallothionein as efficiently as the liver in response to cadmium exposure, resulting in an accumulation of non-metallothionein cadmium in the kidney but not in the liver.

Nomiyama et al. (1982a) studied the concentrations of total cadmium, metallothionein-cadmium, and non-metallothionein-cadmium in the renal cortex of monkeys fed diets containing 30 mg cadmium/kg food for 12 months. They found that the concentration of non-metallothionein-cadmium increased with dose of cadmium to about 35 mg/kg tissue and total cadmium to about 200 mg/kg tissue when the total dose of cadmium was 0.4 g. Similar measurements were made in rabbits given cadmium chloride (0.5 mg cadmium/kg body weight) subcutaneously every day for 21 weeks. There were parallel increases of total cadmium and non-metallothionein-bound cadmium during the initial 4 weeks of dosing; these remained unchanged until the 14th week when striking renal dysfunction appeared. At that time, total cadmium and non-metallothionein-bound cadmium levels fell despite the continued administration of cadmium (Nomiyama & Nomiyama, 1982). Increased knowledge of the intracellular binding or speciation of non-metallothionein-bound cadmium should improve our understanding of the relative roles of metallothionein-bound and non-metallothionein-bound cadmium.

When Cherian et al. (1978) exposed mice by injection and feeding to both cadmium chloride and cadmium-metallothionein, both compounds were absorbed and distributed in the body. However, in the short term, cadmium-metallothionein was selectively accumulated in the kidney and cadmium chloride in the liver.

Most cadmium in human urine is bound to metallothionein (Tohyama et al., 1981b), and good correlation has been found between the urinary cadmium and metallothionein concentrations both in elderly women exposed in the general environment (Tohyama et al., 1981b) and in male cadmium workers (Nordberg et al., 1982; Roels et al., 1983b). Measurement of urinary metallothionein thus provides a good indication of the urinary cadmium level and offers the advantage over cadmium analysis of avoiding the possibility of external contamination. Women were found to have much higher urinary metallothionein concentrations than men, even at similar cadmium levels.

## 6.9 Conclusions

Data from experimental animals and humans have shown that pulmonary cadmium absorption is greater than gastrointestinal absorption. Depending on chemical speciation, particle size, and solubility in biological fluids, up to 50% of the inhaled cadmium compound may be absorbed. The gastrointestinal absorption of cadmium is influenced by the type of diet and nutritional status, iron status appearing to be particularly important. On average, 5% of the total oral intake of cadmium is absorbed, but individual values range from less than 1% to more than 20%. Cadmium may also be transported to the fetus. However, although cadmium accumulates in the placenta, little is transferred to the fetus.

Cadmium absorbed from the lungs or the gastrointestinal tract is stored principally in the liver and kidneys where more than half of the body burden is deposited. Highest cadmium concentrations are generally found in the renal cortex, but as exposure levels increase, a greater proportion of the absorbed cadmium is stored in the liver. The cadmium excretion rate is normally low, and the biological half-time is very long (decades) in the kidneys, muscles, liver, and total body of humans. The cadmium concentrations in most tissues increase with age. In exposed people with renal damage, urinary excretion of cadmium increases and, thus, the whole body half-time is shortened. The renal damage leads to losses of cadmium from the kidney, and the renal concentrations are eventually lower than in people with similar exposure but without renal damage.

Metallothionein is an important transport and storage protein for cadmium and other metals. Cadmium can induce metallothionein synthesis in many organs including the liver and kidney. The binding of intracellular cadmium to metallothionein in tissues protects against cadmium toxicity. Non-metallothionein-bound cadmium may, therefore, have a role in the pathogenesis of cadmium-related tissue injury. The speciation of other cadmium complexes in tissues or biological fluids is unknown.

Urinary excretion of cadmium is related to body burden, recent exposure, and renal damage. In people with low exposures, cadmium in urine is mainly related to body burden. Cadmium-exposed people with proteinuria generally exhibit greater cadmium excretion than such people without proteinuria. After high exposure ceases, urinary cadmium decreases even though renal damage persists. The interpretation of urinary cadmium is thus dependent on a number of factors. The magnitude of gastrointestinal excretion is similar to that of urinary excretion, but it cannot be easily measured. Other excretory routes such as lactation, sweating or placental transfer are insignificant.

Cadmium in faeces is a good indicator of recent daily intake from food in the absence of inhalation exposure. Cadmium in blood occurs mainly in the blood cells, and the plasma concentrations are very low. There are at least two blood compartments, one being related to recent exposure with a half-time of about 2-3 months, and the other probably related to body burden with a half-time of several years.

# 7. EFFECTS ON LABORATORY MAMMALS AND *IN VITRO* TEST SYSTEMS

## 7.1 Single exposure

### 7.1.1 *Lethal dose and lethal effects*

$LD_{50}$ inhalation values are in the range of 500 to 15 000 $mg/m^3$.min for different species (Barrett et al., 1947; Harrison et al., 1947; Hadley et al., 1979). The cause of death is pulmonary oedema.

The $LD_{50}$ after the injection of soluble cadmium compounds is in the range of 2.5-25 mg/kg body weight (Friberg, 1950; Eybl & Sykora, 1966; Commission of the European Communities, 1978). Shortly after large doses are injected, severe endothelial damage is seen in the small vessels of the peripheral nervous system (Gabbiani, 1966) and in the testis (Parizek, 1957). If the animal survives for some hours, the most pronounced changes are found in the liver (Dudley et al., 1982), and liver damage is probably the lethal effect of a single high parenteral exposure.

For most cadmium compounds, the $LD_{50}$ after oral administration is about 10-20 times higher than after parenteral administration, and the readily soluble compounds have a lower $LD_{50}$ values than the insoluble ones (Table 11).

Nomiyama et al. (1978b) showed that the $LD_{50}$ in mice was lower at cold temperatures (+8 °C) than at higher temperatures (+22 or +37 °C), both after oral and peritoneal exposure.

### 7.1.2 *Pathological changes affecting specific systems in the body*

The chronic effects of long-term exposure to low doses of cadmium constitute the main problem for non-occupationally exposed humans. Therefore, the effects of single exposure in animals will be dealt with only briefly, and the main emphasis will be on chronic effects.

Specific effects from a single high dose of cadmium have been described by several investigators and have been reviewed by Friberg et al. (1974, 1986), Commission of the European Communities (1978), and Kawai (1978). One of the most pronounced effects seen was in the gonads (testis and ovary).

Table 11. $LD_{50}$ values for cadmium compounds given to mice and rats by intragastric administration

| Species | Compound | Molecular formula | Relative molecular mass[a] | $LD_{50}$ with confidence limits (mg/kg)[b] | $LD_{50}$ for cadmium ion alone (mg/kg) |
|---|---|---|---|---|---|
| Mouse | cadmium (element) | Cd | 109 | 890 (636-1246) | 890 |
| | cadmium oxide | CdO | 128.4 | 72 (41-113) | 63 |
| | cadmium sulfate | $CdSO_4$ | 208.5 | 88 (69.8-100.2) | 47 |
| | cadmium chloride | $CdCl_2$ | 183.3 | 93.7 (75.5-111.9) | 57 |
| | cadmium nitrate | $Cd(NO_3)_2$ | 236.4 | 100 (78.7-121.8) | 48 |
| | cadmium iodide | $CdI_2$ | 366.2 | 166 (139-193) | 51 |
| | cadmium caprylate | $Cd(C_7H_{15}COO)_2$ | 394.8 | 300 (196-459) | 85 |
| | cadmium carbonate | $CdCO_3$ | 169 | 310 (215-404) | 202 |
| | cadmium stearate | $Cd(C_{17}H_{35}COO)_2$ | 679.4 | 590 (556-624) | 98 |
| | cadmium sulfide | CdS | 144.5 | 1166 (1135-1197) | 907 |
| | cadmium sulfoselenide | CdSe.CdS | 335.8 | 2425 (2393-2457) | 1623 |
| | barium-cadmium stearate | $BaCd(C_{17}H_{35}COO)_4$ | 1383.7 | 3171 (2763-3579) | 258 |
| Rat | cadmium caprylate | $Cd(C_7H_{15}COO)_2$ | 394.8 | 950 (613-1472) | 270 |
| | cadmium stearate | $Cd(C_{17}H_{35}COO)_2$ | 679.4 | 1225 (875-1574) | 203 |
| | barium-cadmium stearate | $BaCd(C_{17}H_{35}COO)_4$ | 1383.7 | 1980 (1736-2224) | 161 |

[a] From: Weast (1974)
[b] From: Tarasenko et al. (1974), Vorobjeva & Sabalina (1975), and Vorobjeva & Bubnova (1981)

### 7.1.2.1 Acute effects on testes and ovaries

Testicular necrosis occurs in experimental animals given single injections of salts corresponding to 2-4 mg cadmium/kg body weight (Parizek & Zahor, 1956; Parizek, 1957). At a later stage, Leydig cells regenerate (Parizek, 1957, 1960; Allanson & Deanesly, 1962). Gabbiani et al. (1974) detected dilation of interendothelial clefts in the small blood vessels of the testis as early as 15 min after an intravenous injection of cadmium salts. Effects on the testis have been extensively reviewed by Gunn & Gould (1970).

The marked effects on the testis after cadmium injection are probably the result of endothelial damage. In the small vessels this damage gives rise to increased capillary permeability. This leads to vascular escape of fluids and blood plasma substances into the interstitium, which results in oedema, decreased capillary blood flow, ischaemia, and testicular cell necrosis (Aoki & Hoffer, 1978; Francavilla et al., 1981).

A single injection of cadmium salts at a dose that induces testicular haemorrhagic necrosis has been shown to induce haemorrhages and necroses in the ovaries of prepubertal rats (Kar et al., 1959), and in the ovaries of adult rats in persistent oestrus (Parizek et al., 1968a). The effect of cadmium on the testis was not dependent on the presence of the hypophysis (Parizek, 1960). Ovarian effects can be prevented by the administration of PMSG hormones (Parizek et al., 1968a). Numerous studies on the effects of cadmium on the testes and other reproductive organs were reviewed by Barlow & Sullivan (1982).

### 7.1.2.2 Acute effects on other organs

A single inhalation exposure to cadmium at concentrations of 5-20 mg/m$^3$ for 50-120 min gives rise to pulmonary oedema in rats and rabbits (Hayes et al., 1976; Bouley et al., 1977; Bus et al., 1978; Dervan & Hayes, 1979; Boisset & Boudene, 1981; Fukuhara et al., 1981). The morphological changes seen in the lung have been described in detail by Strauss et al. (1976).

After the parenteral administration of cadmium at dose levels similar to the $LD_{50}$, pronounced effects were seen in the small blood vessels of, for instance, the nervous system (Gabbiani et al., 1974). Hoffman et al. (1975) noted profound morphological effects in the liver of rats given 6 mg cadmium/kg body weight, and Dudley et al. (1982), examining liver effects from a single injection of cadmium (3.9 mg/kg body weight), concluded that liver was the major target organ in rats for acute cadmium toxicity. Changes in blood pressure shortly after the acute administration of cadmium have also been recorded (Dalhamn & Friberg, 1954; Perry et al., 1970).

Oral administration of cadmium compounds induces epithelial desquamation and necrosis of the gastric and intestinal mucosa, together with dystrophic changes of the liver, heart, and kidneys (Tarasenko et al., 1974; Vorobjeva & Sabalina, 1975).

## 7.2 Repeated and/or long-term exposure

### 7.2.1 Effects on the kidneys

Since the kidney is the critical organ in humans exposed for long periods to relatively small amounts of cadmium (section 8.2.1), results from relevant animal studies will be dealt with in some detail. Even though it is difficult to extrapolate quantitative information from the findings in animals, experiments have provided valuable information concerning mechanisms of cadmium-induced nephropathy and the significance of various biological indicators of exposure and effect, and have supported the findings in humans. For example, Friberg (1950) verified in animal experiments that exposure to cadmium caused a type of proteinuria similar to the one he had found in exposed workers.

Animal studies that have given data on renal effects as well as the corresponding renal cadmium concentrations are summarized in Table 12. An evaluation of organ dose-effect and dose-response relationships is included in section 7.2.1.4.

#### 7.2.1.1 Oral route

Renal lesions were first reported by Prodan (1932) and Wilson et al. (1941) after cats and rats were given large oral doses of cadmium for several months. Prodan (1932) reported varying degrees of desquamation in proximal tubular epithelium (and no changes in the glomeruli) after feeding cats 100 mg cadmium per day for one month. Wilson et al. (1941) reported slight tubular changes in rats after they were exposed for 3 months to a diet containing 62 mg cadmium/kg.

Studies utilizing high exposures have also been performed by Stowe et al. (1972). Ten rabbits received cadmium in drinking-

Table 12. Summary of animal studies with data on both renal cadmium levels and effects

| Species | Route of administration | Exposure level | Duration (months) | Average cadmium level in kidney cortex (mg/kg wet weight) | Renal changes | Reference |
|---|---|---|---|---|---|---|
| Mouse | subcutaneous | 0.25 mg/kg body weight | 6 | 110-170[a] | no effects | Nordberg & Piscator (1972) |
| Mouse | subcutaneous | 0.5 mg/kg body weight | 6 | 170[a] | tubular protein patterns in urine | Nordberg & Piscator (1972) |
| Rat | intraperitoneal | 0.75 mg/kg body weight | 3 | 200[a] | no effects | Bonnell et al. (1960) |
| Rat | intraperitoneal | 0.75 mg/kg body weight | 4 | 300[a] | histological changes in 60% of animals | Bonnell et al. (1960) |
| Rat | subcutaneous | 0.65 mg/kg body weight | 3 | 200 | histological changes | Goyer et al. (1984) |
| Rat | water | 10 mg/litre | 8.5 | 11[a] | no histological changes | Kawai et al. (1976) |
| Rat | water | 50 mg/litre | 8.5 | 35[a] | slight histological changes | Kawai et al. (1976) |
| Rat | water | 100 mg/litre | 8.5 | 90[a] | histological changes | Kawai et al. (1976) |
| Rat | water | 200 mg/litre | 8.5 | 145[a] | histological changes | Kawai et al. (1976) |
| Rat | water | 200 mg/litre | 11 | 200 | total proteinuria and low molecular weight proteinuria | Bernard et al. (1981) |

Table 12 (contd).

| Species | Route of administration | Exposure level | Duration (months) | Average cadmium level in kidney cortex (mg/kg wet weight) | Renal changes | Reference |
|---|---|---|---|---|---|---|
| Rat | water | 50 mg/litre | 3 | 100[a] | decreased insulin and PAH clearance; histological changes | Kawamura et al. (1978) |
| Rat | water | 50 mg/litre | 2.5 | 235 | slight histological changes in proximal tubules | Axelsson & Piscator (1966a); Axelsson et al. (1968) |
| Rabbit | subcutaneous | 0.25 mg/kg body weight | 2.5 | 235 | slight histological changes in proximal tubules | Axelsson & Piscator (1966a); Axelsson et al. (1968) |
| Rabbit | subcutaneous | 0.25 mg/kg body weight | 4 | 460 | more severe histological changes; reduction of alkaline phosphatase activity in renal cortex; total proteinuria | Axelsson & Piscator (1966a); Axelsson et al. (1968) |
| Rabbit | subcutaneous | 0.5 mg/kg body weight | 2.5 | 300 | total proteinuria | Nomiyama et al. (1982b) |
| Rabbit | subcutaneous | 0.5 mg/kg body weight | 0.7 | 200 | proteinuria, glucosuria, and aminoaciduria; decrease in $C_{IN}$ and $Tm_{PAH}$ | Nomiyama & Nomiyama (1982) |
| Rabbit | subcutaneous | 0.5 mg/kg body weight | 1 | 120 | $\beta_2$-microglobulin | Nomiyama et al. (1982b) |
| Rabbit | subcutaneous | 1.5 mg/kg body weight | 1 | 50-200 | decreased tubular readsorption | Nomiyama (1973a); Nomiyama et al. (1978a) |

Table 12 (contd).

| | | | | | | |
|---|---|---|---|---|---|---|
| Rabbit | subcutaneous | 0.5 mg/kg body weight | 2 | 160[a] | slight histological changes | Kawai et al. (1976) |
| Rabbit | water | 160 mg/litre | 6 | 170[a] | extensive fibrosis; pronounced changes | Stowe et al. (1972) |
| Rabbit | water | 50 mg/litre | 10 | 58 | slight tubular atrophy | Kawai et al. (1976) |
| Rabbit | water | 200 mg/litre | 10 | 200 | severe interstitial and tubular fibrosis | Kawai et al. (1976) |
| Rabbit | diet | 300 mg/kg | 4 | 200 | aminoaciduria, enzymuria | Nomiyama et al. (1975) |
| Rabbit | diet | 300 mg/kg | 10 | 300 | proteinuria, glucosuria | Nomiyama et al. (1975) |
| Pig | diet | 50-350 mg/kg | | < 12[a] | equimolar increase in zinc in kidney | Cousins et al. (1973) |
| Pig | diet | 50-350 mg/kg | | 78[a] | decrease in renal leucine aminopeptidase | Cousins et al. (1973) |
| Horse | diet | no cadmium added | lifelong (up to 240) | 75 | renal tubular interstitial changes and fibrosis in 25% of animals | Elinder et al. (1981a,b) |
| Bird | subcutaneous | 0.16 mg/kg body weight | 1.5 | 20[b] | histological changes | Nicholson & Osborn (1983) |

[a] These values are whole kidney concentrations; about 0.8 times kidney cortex values, on average.
[b] Denotes concentrations (wet weight) calculated as 0.2 times dry weight concentrations.

water (160 mg/litre) for 6 months. Kidney function was not investigated, but histopathological examination revealed pronounced morphological changes in the proximal tubules. The mean renal concentration of cadmium was 170 mg/kg wet weight, which would correspond to about 210 mg/kg wet weight in the renal cortex (section 6.4). Still higher doses (300 mg/kg diet) were given to rabbits for 54 weeks by Nomiyama et al. (1975), who observed aminoaciduria and enzymuria after 16 weeks. At this stage, the cadmium concentration in the renal cortex was 200 mg/kg wet weight. Proteinuria and glycosuria appeared at a later stage, 37 and 42 weeks, respectively, after exposure had started. The cadmium concentration in the renal cortex was 300 mg/kg wet weight after 40 weeks.

Morphological changes of the renal tubules were reported by Kawai et al. (1976) in rats given 50 mg cadmium/litre drinking-water for 8.5 months. The average renal cadmium concentration was about 38 mg/kg wet weight which corresponds to about 50 mg/kg in the renal cortex.

Histological lesions in the proximal renal tubules were also found in rats exposed to 200 mg cadmium/litre for 2 months (Itokawa et al., 1978). Histochemical examination of the kidney showed that the proximal tubular epithelium had particularly high cadmium concentrations. The average renal concentrations were 48 mg/kg and 80 mg/kg, respectively, in rats with sufficient and deficient calcium intakes. These levels would correspond to about 60 and 100 mg cadmium/kg in the renal cortex (section 6.4). Inulin clearance was reduced to about a third of the control values in the cadmium-exposed groups, indicating considerable functional damage to the glomeruli. The only reported change in renal tubular function was that the fractional excretion of calcium was increased about 50% in the cadmium-exposed groups.

In a study of 50 rats exposed to 200 mg cadmium/litre in drinking-water for up to 11 months (Bernard et al., 1981), there was an increased prevalence of total proteinuria in the 8th month, when the average cadmium concentration in the renal cortex was about 200 mg/kg.

Kajikawa et al. (1981) also reported morphological changes in the kidneys of rats given drinking-water containing 200 mg cadmium chloride/litre for 91 weeks. Histologically, they found degenerative changes in the proximal convoluted tubules and, using electron microscopy, proliferation of smooth endoplasmic reticulum, vacuolization, and coagulative necrosis of the tubular cells. No significant changes were observed in the glomeruli or interstitial tissue.

When Cousins et al. (1973) gave large amounts of cadmium chloride to pigs (50, 150, 450, and 1350 mg/kg diet), there was a decrease in the activity of leucine aminopeptidase in the kidney cortex at a renal cadmium concentration of 78 mg/kg wet weight, corresponding to a renal cortex concentration of about 100 mg/kg wet weight (section 6.4).

An extensive data base on the renal effects of cadmium in monkeys has been developed in Japan. Several of these studies are summarized in Table 13.

Study I (Nomiyama et al., 1979) was carried out using ten male rhesus monkeys (three years of age). The monkeys were given 100 g of solid feed containing 0, 3, 30, or 300 mg cadmium/kg daily for 37 weeks, followed by 130 g of feed for 18 weeks. Even the solid feed given to the control group contained cadmium at a concentration of 0.13 mg/kg.

In study II, Nomiyama et al. (1987) used 36 male rhesus monkeys (three years of age) and gave them 100 g of solid feed containing 0, 3, 10, 30, or 100 mg cadmium/kg daily for 52 weeks. During the following 52 weeks 150 g was given, and then for the remaining 358 weeks 200 g was given. The solid feed given to the control group contained cadmium at 0.27 mg/kg and zinc at 30 mg/kg.

Study III (Nomura et al., 1988) was performed with 40 female rhesus monkeys given 150 g of solid feed for nine years (Table 14).

In study IV, Nomiyama & Nomiyama (1988) used nine male crab-eating monkeys. Two of the animals were used as controls, and three were given a diet containing cadmium concentrations of 3 mg/kg (190 μg/day) as cadmium chloride (the pelleted food also contained (30 mg zinc/kg)). The remaining four were fed 80 μg cadmium/day in the form of contaminated rice.

In studies I and II, monkeys that had been given feed containing cadmium at 300 mg/kg and 100 mg/kg showed indications of renal dysfunction, such as proteinuria, glucosuria, and aminoaciduria, after 15-16 and 48-91 weeks, respectively. The

Table 13. Renal effects of cadmium in monkeys[a]

| Study | Duration (weeks) | Sex | No. of monkeys | Exposure level (mg/kg diet) | Average cadmium level in renal cortex (mg/kg) | Renal effects (timing, in weeks, of effects) | Other effects (timing, in weeks, of effects) |
|---|---|---|---|---|---|---|---|
| I | 55 | male | 2 | 0 | 163 | | |
| | | | 2 | 3 | 202 | no biological effects | no biological effects |
| | | | 3 | 30 | 596 | no biological effects | no biological effects |
| | | | 3 | 300 | 380[b] | renal dysfunction (15-16) | hepatic dysfunction (12-54) |
| | | | | | 757[b] | $\beta_2$-microglobulinuria (30) | slight anaemia (20) |
| II | 462 | male | 6 | 0 | 328 | | |
| | | | 8 | 3 | 700 | no biological effects | no biological effects |
| | | | 8 | 10 | 1070 | no biological effects | erythrocytopenia (360) |
| | | | 8 | 30 | 1170[b] | renal dysfunction (300-306) $\beta_2$-microglobulinuria (311) | erythrocytopenia (240) |
| | | | 6 | 100 | 635[b] | renal dysfunction (48-91) $\beta_2$-microglobulinuria (138) | erythrocytopenia (120) depressed age-related increase in blood pressure (80) |
| III | See Table 14. | | | | | | |
| IV | 308 | male | 2 | 0 | 18 | | |
| | | | 3 | 3 | 570 | no biological effects | no biological effects |
| | | | 4 | contaminated rice (1.33 mg/kg) | 230 | no biological effects | no biological effects |

[a] From: Nomiyama et al. (1979), Nomiyama et al. (1987), Nomura et al. (1988), Nomiyama & Nomiyama (1988)
[b] The numbers with asterisks are the critical concentrations of cadmium in the renal cortex.

Table 14. Bone and renal effects of cadmium in female monkeys[a]

| Group | No. of monkeys | Exposure level (mg/kg) | Low protein, calcium and phosphorus diet[b] | Low vitamin D diet[c] | Average renal cortex cadmium level (mg/kg) | Renal effects | Bone effects[d] |
|---|---|---|---|---|---|---|---|
| 1 | 5 | 0 | - | - | 58 | no biological effects | no biological effects |
| 2 | 4 | 0 | + | - | | no biological effects | slightly disturbed calcification (after 154 weeks) |
| 3 | 4 | 0 | - | + | | no biological effects | low plasma vitamin $D_3$ |
| 4 | 4 | 0 | + | + | | no biological effects | osteomalacic change (after 77 weeks) reversible by vitamin $D_3$ |
| 5 | 5 | 30[e] | - | - | 1511 | no biological effects | no biological effects |
| 6 | 4 | 30[e] | + | - | | $\beta_2$-microglobulinuria[f] (2000 to 12 000 $\mu$g/day, 67%) | disturbed calcification (after 154 weeks) |
| 7 | 4 | 30[e] | - | + | | no biological effects | low plasma vitamin $D_3$ |
| 8 | 10 | 30[e] | + | + | | $\beta_2$-microglobulinuria[g] (up to 2000 $\mu$g/day) | osteomalacic change (after 77 weeks) reversible by vitamin $D_3$ |

[a] From: Nomura et al. (1988); duration of experiment was 463 weeks (9 years)
[b] 14% protein instead of 20%; 0.3% calcium instead of 0.9%; 0.3% phosphorus instead of 0.9%
[c] No vitamin $D_3$ was added (240 IU was added to the normal diet)
[d] In Group 5 to 8, as depressed age-related increase in blood pressure was seen after 103 weeks of treatment.
[e] 3 mg/kg for the first 52 weeks
[f] Non-progressive lesion, reversibility uncertain; renal effect noted after 193 weeks
[g] Reversible by normal diet and vitamin $D_3$ treatment; renal effect noted after 154 weeks

appearance of increased $\beta_2$-microglobulin was delayed until the 30th and 138th weeks, respectively. However, no definite disturbance of proximal renal tubular function, such as reduced tubular reabsorption of phosphorus, hypophosphataemia or acidosis, was noted during the one-year follow-up. The dose-effect relationship for renal dysfunction was similar to those which have been observed in rabbits and rats, and thus the hypothesis that the susceptibility of monkeys to cadmium may be exceptionally low was not corroborated.

In study II, the group of monkeys given feed containing 30 mg/kg developed urine findings (e.g., proteinuria, glucosuria, aminoaciduria) indicative of renal dysfunction in the sixth year. Postmortem examination revealed degeneration of the proximal renal tubules, but there was no reduction in tubular reabsorption of phosphorus. When the administration of cadmium was discontinued in the fifth year, no abnormality of renal function developed during the follow-up period of four years.

The monkeys in study III (Table 14) given the low nutrition feed plus 30 mg cadmium/kg developed renal function abnormalities after the fourth year. In addition to reduced phenolsulfonphthalein (PSP) clearance and a variable increase in urinary $\beta_2$-microglobulin concentration, many of the monkeys showed mild degenerative changes of the proximal tubular epithelia, but there was no decrease in the tubular reabsorption of phosphorus. However, elevated urinary $\beta_2$-microglobulin did not progress further with continued administration of cadmium. Mild degenerative changes of the proximal tubular epithelia were also noted in the groups of monkeys that had been given a normal diet, low vitamin D diet or low nutrition plus low vitamin D diet, each supplemented with 30 mg cadmium/kg. However, the elevated urinary $\beta_2$-microglobulin level soon returned to normal in those animals fed a normal diet, regardless of continued cadmium administration. This may indicate that the elevated urine $\beta_2$-microglobulin in this group was not caused solely by cadmium exposure.

In study II, some of the monkeys given feed containing 3 mg/kg or 10 mg/kg of cadmium showed cadmium concentrations in the renal cortex as high as 760 mg/kg and 1070 mg/kg, respectively. However, no effect upon renal function was observed during the nine-year period, nor was there any increase in urinary $\beta_2$-microglobulin concentration.

The above data suggest that mild renal dysfunction (proteinuria, glucosuria, and aminoaciduria, but no decrease in the tubular reabsorption of phosphorus) was produced in monkeys exposed to high concentrations of cadmium (30 mg/kg diet or more). It seems, however, that no effects on renal function occur with low-level exposure (10 mg/kg or less). The development of renal dysfunction is assumed to depend upon the amount of cadmium absorbed per day rather than the total amount absorbed in the body.

In study IV, the urinary cadmium level occasionally exceeded 10 μg/litre, but no clinical chemistry changes were reported. Cadmium concentrations in the renal cortex increased proportionally to the dose level and duration of exposure, reaching an average of 450 mg/kg in the group given cadmium chloride and 290 mg/kg in the group fed contaminated rice. This suggests that the chemical form of cadmium does not affect the severity of health effects.

#### 7.2.1.2 *Respiratory route*

Princi & Geever (1950) could find no evidence of renal morphological changes in the kidney of dogs after prolonged inhalation exposure (up to one year) to cadmium oxide or cadmium sulfide dust (average concentration of 4 $mg/m^3$). Routine analysis was performed, but neither the methods used nor the results obtained were described. Friberg (1950) exposed rabbits for about 8 months (3 h per day, about 20 days per month) to cadmium oxide dust with an average concentration of about 8 $mg/m^3$. After 4 months of exposure, moderate proteinuria was detected by the trichloroacetic acid test. Histological examination of the kidneys after 8 months revealed interstitial infiltration of leucocytes in the majority of the exposed rabbits; this was not found in the control group.

#### 7.2.1.3 *Injection route*

Friberg (1950) detected proteinuria in rabbits given subcutaneous injections of cadmium sulfate (0.65 mg cadmium/kg body weight) 6 days per week. Electrophoretic analysis of urine proteins revealed that the proteinuria differed from that caused by injections of uranium salts. More recently, many studies utilizing parenteral administration (with doses generally in the range of 0.25-1.5 mg/kg body weight), different routes of exposure (subcutaneous and intraperitoneal), and a duration of 1-12 months have been performed in mice, rats, and rabbits (Table 12). These experiments have confirmed the nephrotoxic effects of cadmium.

When rabbits were exposed for 16 weeks by subcutaneous injection of either 0.25 mg or 0.5 mg cadmium/kg body weight 3 times a week, there was a significant increase in urinary $\beta_2$-microglobulin excretion indicative of renal tubular dysfunction in the high-dose group after 7 weeks. There was only a slight increase in the serum $\beta_2$-microglobulin/creatinine ratio. Urinary $\beta_2$-microglobulin levels were not related to serum $\beta_2$-microglobulin levels (Piscator et al., 1981).

Rats dosed intraperitoneally, five days/week with 0.6 mg cadmium/kg body weight, showed no abnormal effects after 5 or 6 weeks when renal cadmium levels reached about 100 mg/kg. However, in renal tubular lining cells an increase in lysosomes, microbodies, and smooth endoplasmic reticulum was noted. After 8 weeks renal cadmium levels had reached about 200 mg/kg of tissue and tissue necrosis was observed. The early changes (with a renal cadmium concentration of up to 100 mg/kg) were considered to be adaptive and possibly reversible, whereas morphological changes after 8 weeks with a renal concentration of 200 mg/kg were considered to be irreversible (Goyer et al., 1984).

Nomiyama et al. (1982a) found that non-metallothionein-bound cadmium increased up to about 35 mg/kg tissue in parallel with total cadmium. At that stage, the total cadmium concentration in the renal cortex was in the range 200-800 mg/kg, the total dose of cadmium having been approximately 1 g.

#### 7.2.1.4 *Pathogenesis of cadmium nephrotoxicity*

Various hypotheses have been proposed to explain the pathogenesis of cadmium nephrotoxicity, particularly the role of the metal-binding protein metallothionein (see section 6.8). This protein is inducible by a number of essential metals (Cherian & Goyer, 1978) and may have as its primary function the intracellular storage of zinc and copper (Panemangalore et al., 1983; Templeton et al., 1985). It is also induced following exposure to cadmium. It is now thought that metallothionein protects against cadmium toxicity and that intracellular cadmium bound to metallothionein is nontoxic (Nordberg, 1971; Goyer et al., 1989). There is considerable support for this hypothesis. Pre-treatment of experimental animals with small doses of cadmium prevents the acute toxic effects of a large dose of cadmium (Nordberg et al.,

1975). Parenteral administration of cadmium-metallothionein causes acute tubular toxic effects in the kidney (Nordberg, 1971; Nordberg et al., 1975; Cherian & Nordberg, 1983). By treatment of animals with repeated doses of cadmium, metallothionein synthesis in the renal cortex can be induced. This prevents against subsequent renal toxicity by parenteral cadmium-metallothionein at dose levels that normally give rise to renal damage (Jin et al., 1987a). Rat renal cortical cells isolated from animals pretreated with cadmium were resistant to normally toxic concentrations of cadmium *in vitro* (Jin et al., 1987b). Similar protective effects were observed in kidney cells pretreated with cadmium *in vitro* (Jin et al., 1987b). Human cells in tissue culture, where metallothionein has been induced by pre-treatment with cadmium, become resistant to previously lethal exposure to cadmium (Glennas & Rugstad, 1984).

With this evidence for the protective role of intracellular metallothionein, several theories have been proposed to explain the nephrotoxicity of cadmium. One hypothesis attributes the nephrotoxicity to that fraction of intracellular cadmium not bound to metallothionein (Nordberg et al., 1975; Nomiyama & Nomiyama, 1982; Squibb et al., 1984). Another hypothesis is that extracellular cadmium bound to metallothionein is toxic (Cherian et al., 1976). Cadmium-metallothionein derived from cadmium-induced synthesis in reticulocytes (Tanaka et al., 1985) or released from liver cells is filtered by the renal glomeruli and reabsorbed by the proximal tubular lining cells where it is catabolized, releasing cadmium ions that cause renal damage (Dudley et al., 1985). This hypothesis is supported by the fact that parenterally administered cadmium-metallothionein is very toxic to renal tubular cells and that the plasma metallothionein level increases with cadmium exposure (Goyer et al., 1984; Shaikh & Hirayama, 1979).

Still another hypothesis is that intracellular cadmium interacts with cell membranes resulting in lipid peroxidation (Stacey et al., 1980) and that cadmium may displace essential metals from metallothionein (Petering et al., 1984), thereby depriving important metalloenzymes of essential metal cofactors.

These hypotheses are not mutually exclusive and the relative significance of each of these mechanisms may differ under particular circumstances of exposure.

Goering et al. (1985) reported the development of calcuria in rats injected with cadmium-metallothionein, using the model described by Squibb et al. (1984). Jin et al. (1987a) confirmed their observations, which suggest that this biological effect may be an early event in the development of renal tubular damage. Data from the above studies further validate the use of the cadmium-metallothionein injection model for studying the mechanisms of cadmium-induced tubular injury, since calcuria is also observed in people with chronic elevated cadmium exposure.

#### 7.2.1.5 *General features of renal effects; dose-effect and dose-response relationships*

The available data show that long-term exposure to cadmium leads to renal tubular lesions with proteinuria, glucosuria, and aminoaciduria, and to histopathological changes (Table 12).

It has been reported that cadmium-induced proteinuria differs from glomerular proteinuria (Friberg, 1950) and involves low molecular weight proteins in particular (Axelsson & Piscator, 1966a; Nomiyama et al., 1982b). Thus, this type of proteinuria resembles the "tubular proteinuria" seen in humans. Microscopic examination reveals typical tubular nephropathy, i.e. atrophy and degeneration of tubular cells, especially proximal tubular cells, and interstitial fibrosis (Bonnell et al., 1960; Axelsson et al., 1968; Kawai et al., 1976).

Electron microscopic changes are characterized by interstitial fibrosis and thickening of the basement membrane of the proximal tubular cells (Kawai et al., 1976). The smooth endoplasmic reticulum is dilated or undergoes proliferation, and there is apical cyst formation (Stowe et al., 1972). An increase in the number of lysosomes and swelling of the mitochondria have also been observed. In addition to tubular findings, there have been reports of pathological changes in 30% of the mesangium cells of the glomeruli of dogs (Murase et al., 1974) and increased thickness of the glomerular basement membrane in rats (Scott et al., 1977).

Investigations into renal function have also revealed substantial changes, mainly in tubular function, e.g., reabsorption of glucose (Axelsson & Piscator, 1966a), whereas the changes in glomerular filtration are relatively small (Axelsson & Piscator, 1966a). Effects have generally been seen at average renal cortex concentrations of 200-300 mg/kg wet weight, but some studies have reported effects at considerably lower concentrations. Histopathological changes in rats, rabbits, horses, and birds have been reported at renal cortex concentrations below 100 mg/kg (Table 12). However, in chronically exposed monkeys, signs of renal tubular changes were reported at around 400-1200 mg cadmium/kg (Table 13).

After renal cortex concentrations of cadmium have reached a level of 200-300 mg/kg wet weight, they level off or decrease. No further increase is seen even with continued exposure (Axelsson & Piscator, 1966a; Nomiyama & Nomiyama, 1976a; Bernard et al., 1981). It has also been shown (Friberg, 1952; Axelsson & Piscator, 1966a; Nordberg & Piscator, 1972; Nomiyama & Nomiyama, 1976a) that urinary excretion of cadmium is low during the initial exposure period but a marked increase in the excretion of cadmium occurs subsequently, which coincides with an increase in protein excretion (Friberg, 1952; Axelsson & Piscator, 1966a; Nordberg & Piscator, 1972) (section 6).

Animal studies have shown that, as the cadmium concentration in the renal cortex increases, the first effects to appear are the histopathological changes in the renal tubular cells. The low molecular weight proteinuria and aminoaciduria develop at somewhat higher cadmium concentrations in the renal cortex and, at even higher concentrations, glucosuria, total proteinuria, and other indications of damaged renal function develop. The diagnosis of these effects depends greatly on the sensitivity of the method for analysing the effect. For instance, a method for analysing low molecular weight proteinuria that can accurately measure levels considered normal (0.1 mg/litre or less) will be able to diagnose proteinuria at an earlier stage than a method with a detection limit of 7 mg/litre.

Most of the studies referred to in Table 12 included no data on the prevalence of renal effects in the animals. The results were given in a qualitative way, stating the average cadmium level in the renal cortex at which effects were seen.

Some dose-response data are available from animal studies. Bernard et al. (1981) produced proteinuria in rats exposed to cadmium in drinking-water (200 mg/litre) for up to 11 months. After 8-9 months, a significant increase in group-average proteinuria was seen, coinciding with a 25% prevalence of increased individual proteinuria. The renal cortex cadmium concentration at that time was about 200 mg cadmium/kg (Bernard et al., 1981). Elinder et al. (1981a) studied horses exposed to cadmium present in their normal food. Histopathological changes in the renal cortex were classified and coded in a blind manner, and the prevalence of different degrees of change was calculated for subgroups of horses with different renal cortex cadmium concentrations. The "background" prevalence was 25-30% and there was an increased prevalence (up to 60-75%) with increased average renal cadmium level. At a renal cortex cadmium concentration of about 75 mg/kg, there was a significant increase in the prevalence of histopathological changes.

### 7.2.2 Effects on the liver

Friberg (1950) demonstrated fibrotic changes in the liver of rabbits exposed to repeated subcutaneous injections of cadmium. Periportal and interlobular collagen deposition was found in the liver of rabbits given 160 mg cadmium/litre in drinking-water for 6 months (Stowe et al., 1972). Liver function tests, however, remained within normal limits. The cadmium concentration in the liver was 188 mg/kg wet weight. Tarasenko et al. (1974) demonstrated by histological techniques that dystrophic changes occur in the liver of rats after repeated intragastric administration of cadmium caprylate in a total dose corresponding to 47 mg cadmium/kg body weight per day. These authors also noted an increased level of lactic acid in the blood serum of the animals. Larionova et al. (1974) detected decreased activity of alanine transaminase in liver tissue and depletion of glycogen in rats given barium cadmium laurate by gavage for 8-10 days at a dosage of 169 mg/kg body weight per day (as the laurate). After intraperitoneal injection of cadmium (up to 1.25 mg/kg body weight for periods of up to 6 weeks), decreased glycogen content and increased daily activity of gluconeogenic enzymes in rat liver were reported by Merali et al. (1974) and Chapatwala et al. (1982).

In long-term studies, rabbits given 300 mg cadmium/kg diet for 54 weeks (Kawai et al. 1976) showed some amyloid deposition in the liver. Studies on rats after exposure for 335 days to 1 mg cadmium/litre in drinking-water (Sporn et al., 1970) revealed changes in liver enzyme activities. Rhesus monkeys exposed to 300 mg cadmium/kg in the diet for 12 weeks (Nomiyama et al., 1979) developed increased levels of plasma enzymes (GOT, GPT, and LDH).

### 7.2.3 Effects on the respiratory system

Interstitial pneumonitis and emphysema were found in rabbits exposed to cadmium iron oxide dust (approximately 8 mg/m$^3$) for 4-8 months (Friberg, 1950) and in rats observed for 4-7 months after a single intratracheal administration of cadmium iron oxide dust (3.5 mg/kg body weight) (Vorobjeva, 1957). However, only very slight pulmonary effects were detected in rabbits and rats exposed to nickel-graphite dust at dose levels several times higher than the concentrations of cadmium iron oxide dust.

Yoshikawa et al. (1975) exposed rats to cadmium oxide fumes (0.1 or 1.0 mg cadmium/$m^3$) for up to 3 months. There were 10 rats in each group, and three of the rats in the high exposure group died after about 7 weeks. Lung fibrosis and the first stage of emphysema were observed at the end of the experiment in the high-dose group. Free macrophage cells in the alveoli were more numerous in both groups, and there was an increased surface tension of the surfactants.

Snider et al. (1973) observed signs of emphysema in rats 10 days after 5-15 daily 1-h periods of exposure to cadmium chloride aerosol (10 mg/$m^3$). Also, long-term exposure to comparatively low air concentrations of cadmium oxide (24-50 $\mu$g cadmium/$m^3$) gave rise to pathological changes in the lungs similar to emphysema as well as to cell proliferation in the bronchi (Prigge, 1978). A long-term study (14 months), in which mice and golden hamsters were exposed to different concentrations of cadmium chloride (30 and 90 $\mu$g/$m^3$), cadmium sulfate (30 and 90 $\mu$g/$m^3$), cadmium sulfide aerosols (90-100 $\mu$g/$m^3$), cadmium oxide fume (10-90 $\mu$g/$m^3$), and dust (10-270 $\mu$g/$m^3$), revealed a significantly increased incidence of alveolar hyperplasia and interstitial fibrosis in most of the exposed groups (Heinrich et al., 1989).

Single intratracheal administration of several cadmium compounds (e.g., oxide, sulfide, carbonate, sulfoselenide, caprylate, stearate, cadmium-barium laurate, and cadmium-barium stearate) in doses from 0.5 mg (as the oxide) to 15 mg (as the sulfide) caused the development, over 6 months, of chronic inflammatory changes, emphysema, and atelectasis leading to fibrosis. Exposure to cadmium oxide and caprylate gave rise to the development of nodules of hyaline connective tissue; these resembled silicotic nodules (Vorobjeva & Sabalina, 1975).

### 7.2.4 Effects on bones and calcium metabolism

Male rats fed a normal diet and exposed to cadmium sulfate by inhalation (3 and 0.3 mg/$m^3$, 4 h daily for 4 months) showed decreased serum and urinary calcium concentrations compare to controls. Female rats similarly exposed to cadmium sulfate (2.8 mg/$m^3$, 3 h/week) during pregnancy showed radiological evidence of osteoporosis in addition to hypocalcaemia (Tarasenko et al., 1975).

In a study by Kogan et al. (1972), cadmium chloride and cadmium sulfate were administered subcutaneously to rats at a daily dose of 1 mg/kg body weight for up to 12 months. At 12 months, X-ray analysis of the bones indicated osteoporosis and osteosclerosis. Subsequent histopathology showed an increase in osteoclasts and a bone structure described by the authors to be indicative of osteomalacia.

Oral administration of cadmium chloride in drinking-water to male rats (1 or 4 μg/kg body weight daily for six months) resulted in changes in calcium metabolism and bone structure characteristic of osteomalacia, which were not observed in the control group. No effects were noted in a group of animals given a daily dose of 0.01 μg/kg body weight (Likutova & Belova, 1987).

Rats given 50 mg cadmium/litre in drinking-water for about 9 months showed reduced calcium and phosphorus absorption from the intestine (Sugawara & Sugawara, 1974). In addition, some of the animals showed histological changes in the duodenal mucosa, a finding also reported in Japanese quail (Richardson & Fox, 1974).

Several mineral balance studies have been made on animals fed cadmium. Simultaneous administration of cadmium with a low-protein, low-calcium diet led to a decrease in the calcium and zinc content of bone (Itokawa et al., 1973). Furthermore, Kobayashi (1974) reported that cadmium feeding led to a negative calcium balance in rats.

The decreased calcium absorption and negative calcium balance in cadmium-exposed rats could result from the inhibitory effects of cadmium on the activation of vitamin D in renal cortical cells (Feldman & Cousins, 1973). The renal conversion of 25-hydroxycholecalciferol to 1,25-dihydroxycholecalciferol has been found to be inhibited by high dietary cadmium exposure in rats fed a normal calcium diet, but this effect was not seen on a low-calcium diet (Lorentzon & Larsson, 1977). The metabolically active form of vitamin D (1,25-dihydroxycholecalciferol) is necessary for the normal absorption of calcium from the intestine. Ando et al. (1981) found that the stimulation of calcium absorption by 1-alpha-hydroxy vitamin $D_3$ was inhibited in rats exposed to cadmium by gastric intubation. Furthermore, the concentration of calcium-binding protein in intestinal mucosa may be decreased by cadmium exposure (Fullmer et al., 1980).

Administration of drinking-water containing 10 mg cadmium per litre to rats fed a normal diet over a 9-month period gave rise to decalcification and cortical atrophy in the skeleton (Kawai et al., 1976). Other workers have also reported effects in the bones of rats following several months exposure to cadmium in drinking-water (Itokawa et al., 1974; Kawamura et al., 1978) and in the diet (Takashima et al., 1980; Nogawa et al., 1981a), and following subcutaneous injection (Nogawa et al., 1981a). In these studies, the bones were reported to show more or less severe osteoporosis and osteomalacia. On the other hand, Kajikawa et al. (1981) did not find either osteoporosis or osteomalacia after rats were exposed for 2 years to cadmium in drinking-water (200 mg/litre).

Anderson & Danylchuk (1979) found that exposure of Beagle dogs for six months to cadmium (25 mg/litre in drinking-water) reduced bone turnover rate, a metabolic abnormality consistent with calcium deficiency or osteomalacia. Kawashima et al. (1988) found that feeding a cadmium-contaminated rice diet or a diet containing 3 mg cadmium chloride/kg for six years to crab-eating monkeys did not produce any change in vitamin D metabolism, and there was no evidence of renal dysfunction. In another series of experiments rhesus monkeys were fed diets containing 3, 10, 30 or 100 mg cadmium/kg for 9 years. Serum vitamin D metabolites and renal production of vitamin D remained unchanged, but in animals fed 30 or 100 mg cadmium/kg of diet there was slight but not statistically significant depression of renal 25-hydroxy vitamin $D_1$-hydroxylase activity. No skeletal abnormalities were found in any of these animals.

Bhattacharyya et al. (1988) studied the effects of 0, 0.25, 5 or 50 mg cadmium/kg diets on female mice bred for six consecutive 42-day cycles of pregnancy and lactation and on non-pregnant controls. The multiparous mice exposed to 50 mg/kg experienced significant decreases in body weight (3-11%) and femur calcium content (15-27%), and the femur calcium to dry weight ratios decreased by 5-7%. These results were thought by the authors to provide evidence that the combination of cadmium exposure and multiparity has a synergic effect on bone metabolism.

In the Japanese monkey study III (section 7.2.1 and Table 14), osteomalacic changes were found in the low-nutrition plus low-vitamin-D diet group (group 4) after 77 weeks. These effects were not further exacerbated by feeding cadmium (group 8). These changes were found to be reversed by the administration of vitamin D. Renal effects were found in group 8 after 154 weeks. Therefore, the osteomalacia found in group 8 was diagnosed as not being renal osteomalacia.

Most of the findings discussed above indicate a direct effect of cadmium on bone mineralization, possibly related to calcium deficiency, and an indirect effect on calcium absorption via vitamin D hydroxylation, perhaps leading to osteomalacia. The direct effects develop after long-term cadmium exposure, whereas the indirect effect on vitamin D metabolism occurs only when renal damage is seen in the animals. Osteomalacia only occurred in monkeys fed a diet low in protein, phosphorus, calcium, and vitamin D. Cadmium administration did not increase these effects.

### 7.2.5 *Effects on haematopoiesis*

Anaemia is a common finding in animals after both dietary (Wilson et al., 1941) and parenteral (Friberg, 1950) exposure to cadmium. After dietary exposure, decreased haemoglobin concentration (Decker et al., 1958) and decreased haematocrit (PCV) (Prigge et al., 1977) are among the early signs of cadmium toxicity.

Fox & Fry (1970) and Fox et al. (1971) reported that the cadmium-induced anaemia could be prevented by simultaneous feeding with iron or ascorbic acid. Decreased gastrointestinal absorption of iron due to cadmium may be one mechanism for this anaemia.

After parenteral exposure, iron administration has a beneficial effect on the anaemia (Friberg, 1955). Berlin & Friberg (1960) showed that cadmium injections caused erythrocyte destruction, but there was no indication of an interference with haemoglobin production. Haemolytic anaemia in rabbits was also reported by Axelsson & Piscator (1966b).

### 7.2.6 *Effects on blood pressure and the cardiovascular system*

Chronic oral administration of cadmium compounds to rats (Perry & Erlanger, 1974) induced statistically significant elevation of blood pressure. However, the systolic pressure changes were much smaller than those previously reported by Schroeder & Vinton (1962) and by Schroeder (1964). Furthermore, a different effect was obtained with lower, as compared with higher, doses of cadmium. Rats given 1, 2.5, or 5 mg cadmium/litre in drinking-water for one year had significantly higher blood pressure values than controls. Six months after the beginning of exposure, a statistically significant increase in blood pressure was also observed in rats given 10 or 25 mg/litre, but the increase was not statistically significant after one year of exposure. In rats receiving 50 mg/litre, a statistically significant increase in systolic blood pressure was observed after 12 months of exposure. As discussed in section 8.2.4, these results may be of basic importance in evaluating data on the possible effects of cadmium on the human cardiovascular system.

Several mechanisms have been postulated (Perry & Erlanger, 1974) to explain the effects of chronic cadmium exposure on the cardiovascular system. Oral adminstration of cadmium doses that induce hypertension (and also parenteral administration of cadmium) was shown to increase circulatory renin activity (Perry & Erlanger, 1973). Injection of cadmium into the renal artery of dogs increased sodium reabsorption by the exposed kidney (Vander, 1962), and repeated intramuscular (Perry et al., 1971) or chronic oral administration of cadmium (Lener & Musil, 1971) was reported to increase sodium retention in the body. By morphometric methods, Fowler et al. (1975) demonstrated effects on the renal blood vessels of rats exposed to various concentrations of cadmium (up to 200 mg/litre in drinking-water) for several weeks. Significantly smaller arteriolar diameters were found in the exposed animals than in the controls.

Perry et al. (1977) studied the influence of exposure duration (from 3 to 24 months) and cadmium dose levels in water (from 0.1 to 10 mg/litre) on the blood pressure of Long-Evans rats. A small increase in blood pressure occurred even at the lowest exposure level after 3 months exposure. The greatest increase in blood pressure (3.2 kPa; 24 mmHg) occurred after 24 months exposure to 1 mg/litre, when the average renal cortex cadmium level was 12 mg/kg wet weight. At higher dose levels, the blood pressure increase was less and, at the highest dosage (10 mg/litre for 24 months), the blood pressure did in fact decrease.

Perry et al. (1976) found hypertensive effects in Sprague-Dawley as well as Long-Evans rats. Petering et al. (1979) reported that male rats were more susceptible to these effects than female rats after exposure via drinking-water, but Ohanian & Iwai (1980) found the opposite for rats exposed parenterally.

In several studies (e.g., Kotsonis & Klaassen, 1977; Whanger, 1979; Fingerle et al., 1982), there was no increase in blood pressure after various cadmium doses were given via drinking-water. The type of diet appears to be crucial for the development of hypertension (Whanger, 1979); it can usually only be produced in rats fed a rye-based diet (Perry & Erlanger, 1982). Nishiyama et al. (1986) postulated that cadmium exposure increases sodium and water retention, which are important factors controlling the development of hypertension.

A detailed review of factors influencing the effects of cadmium on the cardiovascular system was reported by the Task Group on Metal Toxicity (1976), with particular reference to those factors that might modify the dose-effect and dose-response relationships.

Rats exposed to cadmium (5 mg/litre) in drinking-water (Kopp et al., 1980a,b, 1983) developed electrocardiographic and biochemical changes in the myocardium, and impairment of the functional status of the myocardium. These effects could be related to (i) decreased high-energy phosphate storage in the myocardium, (ii) reduced myocardial contractility, or (iii) diminished excitability of the cardiac conduction system. Jamall & Sprowls (1987) found that rats fed a diet supplemented with copper (50 mg/kg), selenium (0.5 mg/kg), and cadmium (50 mg/kg) had marked reductions in heart cytosolic glutathione peroxidase, superoxide desmutase, and catalase. They suggested that heart mitochondria are the site of the cadmium-induced biochemical lesion in the myocardium.

Reviews of all aspects of the cardiovascular effects of cadmium on experimental animals have been reported by Perry & Kopp (1983) and Jamall & Smith (1986).

### 7.2.7 Effects on reproductive organs

Cadmium-induced testicular necrosis (section 7.1.2.1) generally results in permanent infertility (Barlow & Sullivan, 1982). Ramaya & Pomerantzeva (1977) found markedly reduced testis weights 1, 3, and 6 months after mice were administered 4 mg cadmium/kg. The animals were sterile and microscopic examination revealed morphological changes in the testis. Krasovskii et al. (1976) noticed decreased spermatozoa motility and spermatogenesis index in rats continuously exposed via food to 0.5-5.0 mg cadmium/kg body weight. In male mice exposed repeatedly by daily subcutaneous injection of cadmium chloride (0.5 mg/kg per day) for 6 months (Nordberg, 1975), there was a decrease in normal testosterone-dependent proteinuria. Morphological examination of the seminal vesicles revealed a smaller weight and size as well as histological indications of lower secretory activity, this being consistent with decreased testosterone activity in these animals.

### 7.2.8 Other effects

Effects on the immune system have been reported after both chronic and acute cadmium exposure. A decrease in the number of antibody-forming cells in the spleen as well as a decrease in antibody production was seen in mice after long-term exposure to cadmium in drinking-water (Koller et al., 1975). An inhibition of the cell-mediated immune response occurred in mice after repeated intraperitoneal injections (Bozelka & Burkholder, 1982). However, no data are available linking these effects to increased susceptibility to infection or other secondary dysfunctions.

Gestational exposure to cadmium (4.2 and 8.4 μg/ml in drinking-water) results in decreased birth weight, retarded growth, delayed development of the sensory motor coordination reflexes, and increased motor activity. Cadmium exposure during critical periods of development might result in developmental and behavioural deficits with long-term implications for adult behaviour (Mohd et al., 1986).

## 7.3 Fetal toxicity and teratogenicity

In several species of laboratory rodents, large doses of cadmium salts induce severe placental damage and fetal deaths when given at a late stage of pregnancy, and teratogenic effects, such as exencephaly, hydrocephaly, cleft lips and palate, microphthalmia, micrognathia, clubfoot, and dysplastic tail, when given at early stages of gestation.

A single subcutaneous injection of cadmium chloride, acetate, or lactate (4.5 mg cadmium/kg body weight) given to Wistar rats from the 17th to the 21st day of pregnancy (Parizek, 1964) led to the rapid development of severe placental damage in all rats and to fetal death. Placental damage was not dependent on the presence of fetuses, but it was not possible to decide whether fetal lethality resulted from the placental lesion or from a direct effect of cadmium on the fetuses (Parizek, 1964). Similar effects were observed at a dose level of 3.3 mg cadmium/kg body weight

(Parizek et al., 1968b). Placental damage and fetal deaths were also observed after cadmium administration to pregnant Swiss albino mice (Chiquoine, 1965).

Teratogenic effects can be observed when doses close to the $LD_{50}$ (Table 11) for cadmium salts are administered to pregnant females at critical stages of embryogenesis. These effects were demonstrated with intravenous injections of cadmium sulfate in hamsters (Ferm & Carpenter, 1968; Mulvihill et al., 1970; Ferm, 1972) and with intraperitoneal (Barr, 1973), subcutaneous (Chernoff, 1973), or dietary (Scharpf et al., 1972) administration of cadmium chloride in rats. Teratogenic effects induced by cadmium salts have also been demonstrated in mice (Ishizu et al., 1973).

The character of the changes induced is dependent on the species and on the stage of embryogenesis. As little as 123 μg/litre in mouse embryo cultures produced exencephaly apparently by re-opening the closed neural tube (Schmid et al., 1985). Either facial defects or limb abnormalities were induced by cadmium when administered to pregnant hamsters on day 8 or 9 of gestation (Ferm, 1971). Both jaw defects and cleft palate were observed in the offspring of rats given daily subcutaneous cadmium chloride injections (8 mg/kg body weight) on days 13-16 or 14-17 of pregnancy, but cleft palate was not observed when this dosage was given on days 15-18 or 16-19 of pregnancy (Chernoff, 1973). Anophthalmia or microphthalmia and dysplastic ears were induced by approximately 2 mg of cadmium as the chloride given intraperitoneally to pregnant rats on the 9th but not on the 11th day of pregnancy (Barr, 1973). Other effects observed in these studies included decreased lung weight in the offspring of rats subjected to cadmium during pregnancy (Chernoff, 1973) and deficiencies in bone formation and delays in bone ossification (Mulvihill et al., 1970; Scharpf et al., 1972).

The dose-dependent fetal mortality and teratogenicity response was established in studies with subcutaneous administration of cadmium chloride to rats (Chernoff, 1973) and mice (Ishizu et al., 1973). The no-observed-effect level with respect to malformations was found in the latter study to be 0.33 mg/kg body weight.

All the teratogenic effects mentioned above were induced by parenteral administration of very high doses of cadmium salts. However, in a rat study by Scharpf et al. (1972), very high peroral doses (20, 40, 60, or 80 mg/kg body weight given by gavage daily from days 6 to 19 of pregnancy) of cadmium chloride were used with the simultaneous administration of sodium chloride, and internal teratological examinations were performed. Heart and kidney abnormalities were the major internal defects, but their incidence was not directly related to the dose of cadmium chloride administered. At the lowest dose level, heart abnormalities were detected in 19.7% of the 127 fetuses (abnormalities in the control group were seen in 6.6% of 107 fetuses) and teratoma of the kidney was observed in 15.7% of these fetuses.

Cvetkova (1970) exposed pregnant female rats via the respiratory route to cadmium sulfate (2.8 mg/m$^3$, 4 h daily) and, on the 22nd day, killed half of them to examine the embryos. The number of embryos in exposed rats was the same as in a control group, but the mean weight was lower in the exposed group. In the exposed rats, where pregnancies were allowed to proceed to full term, the average weight of the offspring was lower than in the controls both at birth and after 8 months. The rats born to the exposed group also had increased mortality during the first 10 days after birth.

When mice were exposed for several generations to cadmium in drinking-water (10 mg/litre), fetal mortality, runting, and malformations were observed (Schroeder & Mitchener, 1971). The external malformations, consisting of sharp angulation of the distal third of the tail, were observed in 16.1% of 255 offspring ($F_1$ and $F_{2A}$ generation), and 87 deaths before weaning (30.5%) were recorded.

Ferm & Carpenter (1968) showed that zinc injected simultaneously with cadmium could protect against the teratogenic effects of cadmium, and a similar protective action was found for selenium (Holmberg & Ferm, 1969). Maternal zinc deficiency can produce congenital malformations (Hurley et al., 1971). This was confirmed by Parzyck et al. (1978), who also found that intraperitoneal injection of 1.5 mg cadmium/kg body weight to pregnant rats increased the prevalence of malformations. The increase was greater at this cadmium dose than the increase due to zinc deficiency. Combined zinc deficiency and cadmium exposure caused a very high incidence of fetal deaths.

Further experimental data on rats provided by Samarawickrama & Webb (1979) indicate that maternal cadmium exposure gives rise to a fetal zinc deficiency and that this is one cause of the teratogenic effects observed. Intravenous cadmium injections to pregnant rats at doses ranging from 0.25 to 1.25 mg/kg body weight on day 12 of gestation produced a dose-related decrease in fetal uptake of a dose of 65 mg zinc given 4 h later. Maternal cadmium exposure (1.25 mg/kg body weight) was shown to result in decreased activity of a fetal zinc-dependent enzyme thymidine kinase, which is responsible for the incorporation of thymidine in DNA. Additional evidence that cadmium-induced fetotoxicity is related to a cadmium-induced fetal zinc deficiency was reported by Daston (1982), who found that co-administration of zinc (12 mg/kg body weight) almost totally eliminated severe fetal lung lesions when pregnant rats were given cadmium (8 mg/kg body weight) on gestation days 12-15.

## 7.4 Mutagenicity

Studies on Drosophila (Ramel & Friberg, 1971; Vorobjeva & Sabalina, 1975) failed to show any chromosomal abnormalities after exposure to various cadmium compounds. Some *in vitro* studies of cultured human lymphocytes and fibroblasts were also negative (Paton & Allison, 1972; Deknudt & Deminatti, 1978; Kogan et al., 1978). Shiraishi et al. (1972) reported a marked increase in the frequency of chromatid breaks, translocations, and dicentric chromosomes in leucocytes, from one person, cultured in a medium containing 62 mg cadmium/litre (as the sulfate) for 4-8 h.

Andersen et al. (1983) found that the average chromosome length in human lymphocytes was initially reduced when they were cultured in a medium containing cadmium chloride (1.1 mg cadmium/litre), but subsequently returned to normal. This effect was probably related to the synthesis of metallothionein and complexing with cadmium. Watanabe et al. (1979) observed aneuploidy in rat oocytes with cadmium accumulation in the ovary after exposure to cadmium chloride *in vivo*.

Rohr & Bauchinger (1976) found a reduced mitotic index in hamster fibroblasts cultured in 100 $\mu$g cadmium/litre (as the sulfate) and chromosome damage at concentrations above 500 $\mu$g cadmium/litre. Deaven & Campbell (1980) showed that the effects on cultured hamster cells depended on the type of medium used.

There appears to be an acute effect of cadmium following the injection of 0.6-2.8 mg cadmium/kg body weight (Felten, 1979). After 6 h, there was an increased frequency of chromatid breaks in bone marrow cells and chromosome gaps and breaks in spermatocytes, which could be associated with the acute effects on haematopoiesis (section 7.2.5) and on the testis (section 7.1.2.1).

A summary and graphical presentation of the available evidence on genetic and related effects of cadmium in various *in vivo* and *in vitro* test systems has been presented by IARC (1987a). Although prokaryote test systems reveal no effects, variable results have been observed in lower eukaryotes, mammalian cells *in vitro*, and mammals *in vivo*.

## 7.5 Carcinogenicity

Intramuscular or subcutaneous administration of metallic cadmium or cadmium compounds can induce sarcomata at the site of injection. This local effect of cadmium was demonstrated with intramuscular administration of metallic cadmium (cadmium powder in fowl serum) to hooded rats (Heath et al., 1962), subcutaneous administration to Chester Beatty rats of cadmium as the sulfide and oxide (Kazantzis, 1963; Kazantzis & Hanbury, 1966) or sulfate (Haddow et al., 1964), intramuscular injection of cadmium chloride to Wistar rats (Gunn et al., 1967), and subcutaneous injection to Sprague-Dawley rats (Nazari et al., 1967). Transplantability of tumours induced in these studies (Heath & Webb, 1967) and metastases into regional lymph nodes and into lungs (Kazantzis & Hanbury, 1966) were reported. Intratesticular injection of cadmium chloride to White Leghorn cockerels was reported to induce teratoma at the site of injection (Guthrie, 1964).

After Hoffman et al. (1985) injected 1.9 mg cadmium chloride (1.2 mg/kg body weight) directly into the ventral prostatic lobe of 100 12-month-old male rats, simple hyperplasia was found in 38 of the rats, atypical hyperplasia in 29, atypical hyperplasia with severe dysplasia in 11, and invasive prostatic cancer in 5 animals. Hoffman et al. (1988) reported changes in the ultrastructure of prostate epithelial cells in rats injected into the ventral prostate with 2.2 or 3.3 mg cadmium/kg body weight. In animals given oral treatment via the drinking-water (29.9 or 115 mg cadmium/kg body weight), there were changes ranging in severity up to dysplasia but no evidence of carcinoma.

A single parenteral administration of cadmium salts can induce necrosis of the testis (see section 7.1.2.1). After one year, the remnants of the necrotic testis were shown to contain masses of cells showing the typical structure of Leydig cells (Parizek, 1960).

This regeneration of testicular Leydig cells damaged by cadmium can result in Leydig cell neoplasia (Gunn et al., 1963, 1965; Lucis et al., 1972). The ultrastructural features of cadmium-induced Leydig cell tumours correspond in most respects with the fine structural features of normal Leydig cells (Reddy et al., 1973).

Other injection studies and peroral studies did not demonstrate increased malignancy (Schroeder et al., 1964, 1965; Loser, 1980), but the doses were low compared to those necessary to induce renal damage. In one peroral study (Kanisawa & Schroeder, 1969), rats were exposed to 5 mg cadmium/litre in drinking-water for up to 2 years. There were 7 malignant tumours among 47 cadmium-exposed male rats and 2 tumours among 34 male control rats. This indicates a doubling of the tumour rate, but because of the low statistical power of the study, the increase was not statistically significant and the authors concluded that ingestion of these cadmium doses was not carcinogenic.

Some studies have been specially designed to investigate the possible role of cadmium in cancer of the prostate. The prostate gland, like the testis, is of particular interest with respect to cadmium toxicity because these organs contain greater concentrations of zinc than any other tissues and it has been suggested that cadmium may affect prostate growth by competition with zinc (Gunn et al., 1961). Levy et al. (1973) gave three groups of rats weekly subcutaneous injections of cadmium sulfate at concentrations of 0.022, 0.044 or 0.087 mg cadmium per rat (average weight 220 g at the start and 410 g at the end of the 2-year exposure). Weekly injections of water were given to the 75 control rats. The liver cadmium level was 80 mg/kg in the highest-dose group, but no malignant changes were found in the prostate. No difference was seen between exposed and control rats with respect to malignant changes in other organs.

Levy & Clack (1975) and Levy et al. (1975) conducted 2-year studies in rats and mice designed to detect carcinogenic effects in the prostate. The animals in both experiments were given weekly administrations of cadmium sulfate by stomach tube. The rats were given from 0.08 to 0.35 mg/kg body weight and the mice 0.44 to 1.75 mg/kg. Extremely low levels of cadmium were found in the kidney after 2 years (5 mg/kg wet weight in rats), but no macroscopic or microscopic changes were seen in any tissue at these low doses.

In a long-term study on Fisher rats, Sanders & Mahaffey (1984) administered cadmium oxide (25 μg) in single or repeated doses by intratracheal instillations. There was no evidence for pulmonary or prostate carcinogenicity, but increases in mammary tumours and in tumours at multiple sites in male rats were reported.

It has been reported that inhalation of a cadmium aerosol causes lung cancer in Wistar rats (Takenaka et al., 1983). Three groups of 40 rats were continuously exposed to cadmium chloride aerosols for 18 months, the air cadmium concentrations being 12.5, 25, and 50 μg/m$^3$. A control group of 41 rats was also studied. The study was terminated after 31 months, and no lung cancers were seen in the control group. However, in the exposed groups, the incidence was 15%, 53% and 71%, respectively, at increasing exposure levels. Even at these relatively low exposure levels, there was a clear dose-response relationship. Histologically the experimentally induced tumours were adenocarcinomas, epidermoid carcinomas, mucoepidermoid carcinomas, and combined epidermoid and adenocarcinomas.

In a subsequent study, rats were exposed to inhalable aerosols of cadmium sulfate and cadmium oxide and fume and dust at > 30 μg cadmium/m$^3$ and to cadmium sulfide at > 90 μg cadmium per m$^3$ for periods of up to 18 months. Bronchoalveolar benign and malignant adenomas, squamous cell carcinomas, and combined forms developed at high primary tumour rates with all four forms of cadmium tested even after discontinuous exposure for 40 h/week for 6 months. No primary tumour was found with cadmium oxide fume at a concentration of 10 μg cadmium/m$^3$ or cadmium oxide dust (at 30 μg cadmium/m$^3$) when combined with a zinc oxide aerosol (Oldiges et al., 1989).

In a further study, male and female Syrian golden hamsters and female NMRT mice were exposed to cadmium chloride, sulfate, oxide, and sulfide at concentrations of between 10 and 270 μg cadmium/m$^3$. The exposure was continuous (19 h/day, 5 days/week) for 50 to 70 weeks and was followed by a 50-week observation period. No increase in the lung tumour rate was observed in either the mice or hamsters (Heinrich et al., 1989), but in both species exposure to cadmium caused multifocal bronchoalveolar hyperplasia, the extent of which varied with the compound used, its concentration, and the length of exposure. The most severe changes were found after cadmium oxide inhalation (Aufderheide et al., 1990).

A synergistic effect has been shown in rat renal tumours induced by dimethylnitrosamine when followed by cadmium chloride given by intramuscular injection. In this study, cadmium appeared to enhance the initiation of dimethylnitrosamine-induced cancer (Wade, 1987).

## 7.6 Host and dietary factors; interactions with other trace elements

The toxic effects of cadmium in experimental animals have been shown to be dependent on genetic factors, stage of ontogenic development, functional state of the organism, and simultaneous or previous exposure to certain environmental influences, including exposure to certain nutrients.

Resistance to cadmium-induced testicular necrosis is determined by a single autosomal recessive gene (*cdm*) in inbred mice (Taylor et al., 1973). The teratogenic effects of cadmium are dependent on the stage of embryogenesis (section 6.3). The stage of postnatal development of certain organs may be of importance for the toxic effects of cadmium, as has been shown for the testis (Parizek, 1957, 1960), ovaries (Kar et al., 1959), and central nervous system (Gabbiani et al., 1967).

Pretreatment with small, non-toxic doses of cadmium salts has been shown to induce resistance to testicular or lethal effects (Terhaar et al., 1965; Ito & Sawauchi, 1966). The probable mechanism is induction of metallothionein synthesis by the pretreatment. This enables the subsequent dose of cadmium to be bound rapidly to metallothionein, which renders it less acutely toxic (Nordberg, 1971). Similarly, the protective effect of zinc against cadmium toxicity could also be, at least in part, dependent on the induction of an increased synthesis of metallothionein-like proteins (Webb, 1972; Davies et al., 1973).

Selenium compounds are known to be highly effective in preventing the reproductive toxic effects of cadmium (Kar et al., 1960; Mason & Young, 1967; Parizek et al., 1968a,b), lethality to rats (Parizek et al., 1968b) and mice (Gunn et al., 1968), and teratogenicity (Holmberg & Ferm, 1969). Fetal lethality (Parizek et al., 1968b) and teratogenic effects (Holmberg & Ferm, 1969) can be prevented when selenium compounds are given at the same time as cadmium.

Simultaneous administration of mercuric and cadmium compounds has been shown to have an additive effect (Gale, 1973). Oral administration of nitrilotriacetate with large oral doses of cadmium chloride provided protection against the lethality of cadmium and had no potentiating effect on the teratogenicity and fetal accumulation of cadmium (Scharpf et al., 1972). This was confirmed by Engström (1979), who also showed that simultaneous oral exposure to cadmium and sodium tripolyphosphate decreased the mortality expected at the cadmium level used. However, when nitrilotriacetate or sodium tripolyphosphate was given subcutaneously with cadmium, the mortality rates were increased (Engström & Nordberg, 1978; Andersen et al., 1982). The chelation of cadmium in the gastrointestinal tract decreased the uptake of cadmium, whereas parenteral exposure to cadmium and chelating agents caused a higher renal cadmium concentration than cadmium alone.

The interaction of cadmium with certain trace elements can produce symptoms characteristic of trace element deficiencies. As a result, chronic cadmium toxicity in certain animal species closely resembles zinc and/or copper deficiency and can be prevented by administering higher doses of the salts of these trace elements (Petering et al., 1971, 1979; Mills & Delgarno, 1972). Cadmium-calcium interactions are discussed in section 7.2.4.

An increased toxicity of cadmium was reported in animals on low-protein diets (Fitzhugh & Meiller, 1941), this being due partly to rapid intestinal absorption of cadmium (Suzuki et al., 1969). Lack of dietary calcium seems to play a role similar to lack of dietary protein in increasing the toxicity of cadmium (Suzuki et al., 1969). Supplements of dietary ascorbic acid almost completely prevented cadmium-induced anaemia and improved the growth rate (Fox & Fry, 1970).

Ambient temperature (Nomiyama et al., 1978b) and the energy or protein level in the diet have been reported to influence the $LD_{50}$ of cadmium in mice.

Various aspects of the interactions between cadmium and other trace elements were discussed in greater detail by the Task Group on Metal Interactions (1978).

## 7.7 Conclusions

Inhalation exposure at high levels causes lethal pulmonary oedema. Single high-dose injection gives rise to testicular and non-ovulating ovarian necrosis, liver damage, and small vessel injury. Large oral doses damage the gastric and intestinal mucosa.

Long-term inhalation exposure and intratracheal administration give rise to chronic inflammatory changes in the lungs, fibrosis, and appearances suggestive of emphysema. Long-term parenteral or oral administration produces effects primarily on the kidneys, but also on the liver and the haematopoietic, immune, skeletal, and cardiovascular systems. Skeletal effects and hypertension have been induced in certain species under defined conditions. Teratogenic effects and placental damage occur, depending on the relation between the exposure and the stage of gestation, and may involve interactive effects with zinc.

Of greatest relevance to human exposure are the acute inhalation effects on the lung and the chronic effects on the kidney. Following long-term exposure, the kidney is regarded as the critical organ. The effects on the kidney are characterized by tubular dysfunction and cell damage, although glomerular dysfunction may also occur. A consequence of renal tubular dysfunction is a disturbance of calcium and vitamin D metabolism. According to some studies, this has led to osteomalacia and/or osteoporosis, but these effects have not been confirmed by other studies. A direct effect of cadmium on bone mineralization cannot be excluded. The toxic effects of cadmium in experimental animals are influenced by genetic and nutritional factors, interactions with other metals, in particular zinc, and pretreatment with cadmium, which may be related to the induction of metallothionein.

IARC (1976, 1987b) accepted as sufficient the evidence that cadmium chloride, sulfate, sulfide, and oxide can give rise to injection-site sarcomata in the rat and that the chloride and sulfate can induce interstitial cell tumours in the testis of rats and mice, but found oral studies inadequate for evaluation. One recent lifetime study (18 months), in which rats were subjected to continuous inhalation of a cadmium chloride aerosol at low concentration, showed a high incidence of primary lung cancer with evidence of a dose-response relationship. Studies on the genotoxic effects of cadmium have given discordant results, most of the positive results indicating chromosomal effects after short-term high-level exposure.

# 8. EFFECTS ON HUMANS

Most of the available epidemiological studies or group observations, as well as the clinical studies, have been performed either on occupationally exposed workers or on Japanese populations in cadmium-polluted areas. A great deal of epidemiological data has resulted from studies in polluted areas of Japan (Cooperative Research Committee on Itai-itai Disease, 1967; Shigematsu et al., 1978; Japan Cadmium Research Committee, 1989) and, more recently, from smaller studies in other countries (Drasch et al., 1985; Philipp, 1985; Hahn et al., 1987; Roels et al., 1989; Thun et al., 1989; Likutova, 1989). Comprehensive summaries of these studies have also been published (Tsuchiya, 1978; Friberg et al., 1986; Nomiyama, 1986).

Many of these studies have focused on the detection of early signs of kidney dysfunction. Others have investigated clinical signs of disease such as renal stones and pulmonary impairment. Until the middle of the 1970s, particular attention was given in Japan to the detection of and screening for bone disease (e.g., Itai-itai disease). More recently the role of cadmium in human carcinogenesis and mortality has also been studied.

Exposure to cadmium produces a wide variety of effects involving many organs and systems. From the point of view of preventive medicine, the detection of early effects on the kidneys is of particular importance in order to prevent more serious renal effects and those on the lungs or bones. Recent studies indicating that chronic exposure to cadmium may give rise to cancer will be reviewed in some detail.

## 8.1 Acute Effects

### *8.1.1 Inhalation*

Acute cadmium poisoning and, in some cases, death have been reported among workers shortly after exposure to fumes when cadmium metal or cadmium-containing materials have been heated to high temperatures (Beton et al., 1966; Blejer, 1966; Dunphy, 1967). The principal symptom in acute cases, both fatal and non-fatal, is respiratory distress due to chemical pneumonitis and oedema (MacFarland, 1979; Lucas et al., 1980). At an early stage, the symptoms may be confused with those of "metal fume fever".

In working environments where cases of acute poisoning occurred, cadmium concentrations were usually very high. For instance, in one case the fatal air concentration of cadmium oxide fume from a furnace was approximately 50 mg/m$^3$ for a period of about 1 h (a dose of 2900 mg/m$^3$.min) (Barrett & Card, 1947). In another case, the lethal dose was 2600 mg/m$^3$.min (Beton et al., 1966), i.e. a 5-h exposure to 8.6 mg/m$^3$. Friberg et al. (1974) estimated that an 8-h exposure to 5 mg cadmium/m$^3$ may well be lethal.

### 8.1.2 Ingestion

During the period 1940-50, cases of acute food poisoning occurred mainly due to the substitution of cadmium for scarce chromium in the plating of many cooking utensils and containers. Food contamination arose when acid foods and drinks were prepared and stored in contact with cadmium-plated surfaces. Rapid onset with severe nausea, vomiting, and abdominal pain were characteristic symptoms (US Public Health Service, 1942; Cole & Baer, 1944; Lufkin & Hodges, 1944). Effects also occurred following the consumption of drinks with a cadmium concentration of approximately 16 mg/litre from an automatic vending machine in which drinking-water was cooled in a tank constructed with cadmium-containing solder (Nordberg et al., 1973). Recovery from acute poisoning appears to be rapid and complete. The amount of cadmium absorbed is probably very limited due to vomiting and the consequential short presence of cadmium in the gastrointestinal tract. However, no follow-up studies of people who have experienced acute cadmium poisoning have been reported.

## 8.2 Chronic Effects

Lower cadmium concentrations with longer periods of exposure than those described above will cause chronic cadmium poisoning. Fully developed poisoning among industrial workers shows two main effects: renal dysfunction and emphysema (Friberg, 1948a,b, 1950). The kidney is most frequently the critical organ, but under certain conditions (short-term peak exposures) it may be the lung (Bonnell, 1955). For people in the general environment, exposure is usually by the oral route and the kidney is the critical organ.

### 8.2.1 Renal effects and low molecular weight proteinuria

#### 8.2.1.1 In industry

Renal dysfunction is one of the characteristic signs of cadmium poisoning, and many cadmium workers have developed proteinuria, renal glucosuria, and aminoaciduria. In working environments with high cadmium exposure levels, workers have also developed hypercalciuria, phosphaturia, and polyuria (Friberg, 1950; Clarkson & Kench, 1956; Kazantzis et al., 1963; Tsuchiya, 1967; Lauwerys et al., 1974a, 1979b), and some have suffered from renal colic due to recurrent stone formation (Friberg, 1950; Ahlmark et al., 1961; Adams et al., 1969; Scott et al., 1976; Kanzantzis, 1979). The polyuria is due to loss of urinary concentrating ability (Kazantzis, 1979), and, in addition, the kidneys of cadmium-poisoned workers lose their ability to handle an acid load after a standard $NH_4Cl$-loading test. These are signs of distal tubular damage, and in a few severe cases, the renal damage progresses to a reduction in glomerular filtrations (see section 8.2.1.5).

Renal function, as measured by inulin or creatinine clearance and urine concentrating capacity, was depressed in several poisoning cases (Friberg, 1950; Bonnell, 1955; Bernard et al., 1979). Thus, in the more advanced cases, there is a combination of tubular and glomerular effects. In most of the early cases, only proteinuria, mild in comparison with the proteinuria in many other renal disorders, has been reported as a sign of renal dysfunction, and other signs of kidney dysfunction were not evident (Piscator, 1966a,b).

Since Friberg first observed the urinary proteins of cadmium workers (Friberg, 1950), the proteinuria has proved to involve proteins with a molecular weight of 10 000 to 40 000 and is the so-called tubular proteinuria (Butler & Flynn, 1958). Table 15 contains data on proteins in urine useful for the diagnosis of cadmium-induced proteinuria. The increased excretion of low molecular weight proteins in urine from cadmium-exposed workers has been found to apply to $\beta_2$-microglobulin, lysozyme (muramidase), ribonuclease, immunoglobin chains, retinol-binding protein, and $\alpha_1$-microglobulin (Piscator, 1966a; Peterson et al., 1969; Peterson & Berggård, 1971; Lauwerys et al., 1974a; Bernard et al., 1976, 1982b). In groups of exposed and unexposed workers, the urinary $\beta_2$-microglobulin concentrations follow log-normal distributions, and an operational definition for what is an "increased" level should be established for each population studied (Kjellström et al., 1977a).

Table 15. Excretion of urinary proteins in healthy people and in cases of glomerular and tubular disorders

| Protein type | Normal plasma concentration (mg/ml) | Normal filtered amount in primary urine (mg/24 h) | Urinary excretion (mg/24 h) | | | Reference |
|---|---|---|---|---|---|---|
| | | | Healthy people | Glomerular disorders | Tubular disorders | |
| Total protein | | | 43-127 | 310-54 100 | 129-1570 | Peterson et al. (1969) |
| Albumin | 50 | 500 | 3.9-24 | 88-48 800 | 13.8-578 | Mogensen & Solling (1977) |
| Retinol-binding protein | | | 0.11 | | 20-150 | Peterson & Beggard (1971) |
| | | | 0.04 | 3 | 45 | Kanai et al. (1971) |
| $\beta_2$-microglobulin | 0.002 | 300 | 0.06-0.21 | 0.06-4.7 | 1.1-105 | Peterson (1971) |
| | | | 0.073 | | | Evrin et al. (1971) |
| Lysozyme | | | 0-2[a] | | | Prockop & Davidson (1964) |
| | | | 0.07-1.1[a] | | 47-130[a] | Harrison et al. (1968) |
| Ribonuclease | | | 0.24-1.5[a] | | 1.9-10[a] | Harrison et al. (1968) |

[a] Values reported as mg/litre

Proteinuria is known to be an early sign of cadmium poisoning, but the degree of proteinuria varies with time. In a group of 40 workers with heavy exposure to cadmium, it was found that proteinuria was persistent and even increased several years after cessation of exposure, as evaluated by qualitative methods (Friberg & Nystrom, 1952; Piscator, 1966a). Tsuchiya (1976) examined five cadmium-exposed workers who showed proteinuria. Ten years after cessation of exposure, three of them no longer revealed proteinuria, but two of these showed a high urinary $\beta_2$-microglobulin level, as did the two workers with persistent total proteinuria. Four workers in a British pigment factory still had grossly elevated $\beta_2$-microglobulin levels despite removal from exposure many years earlier (Stewart & Hughes, 1980).

According to recent observations using quantitative proteinuria methods (Roels et al., 1982), total proteinuria in 19 workers had not changed 4 years after exposure ceased. In those 11 workers for whom urinary $\beta_2$-microglobulin was measured before and after cessation of exposure, an increase was invariably seen. Eight of the workers had abnormal $\beta_2$-microglobulin levels before exposure ceased, whereas three had normal levels before and developed abnormal levels after cessation. It can be concluded that cadmium-induced tubular proteinuria is irreversible in most workers, at least for several years.

A marked increase in urine cadmium level may reflect cadmium-induced nephropathy if exposure has been chronic and correlates with low molecular weight proteinuria (section 6.5.1.2). Lauwerys et al. (1979b) proposed a biological threshold of 10 μg cadmium/μg urinary creatinine for males occupationally exposed to cadmium. Smith et al. (1980) found that workers with low exposure to airborne cadmium had an average urinary cadmium level of 13.1 μg/litre, whereas workers with long histories of work in areas with substantial airborne cadmium had an average level of 45.7 μg/litre. The high-exposure group showed a significant reduction in urinary clearance and increased $\beta_2$-microglobulin excretion. Buchet et al. (1980) found increased excretion of both low and high molecular weight proteins and tubular enzymes in workers excreting more than 10 μg cadmium/g creatinine or with a blood cadmium level above 10 μg cadmium per litre.

Retinol binding protein (RBP) has been shown to correlate well with $\beta_2$-microglobulin in urine with a pH value greater than

5.5. and is equally sensitive for detection of tubular proteinuria (Bernard et al., 1982a,b). This protein occurs in serum complexed to prealbumin and retinol, but, after retinol is delivered to target cells, RBP rapidly dissociates from prealbumin, is filtered through the glomerulus, and is reabsorbed by the tubule (Peterson, 1971).

Renal tubular brush border enzymes may also be excreted in chronic cadmium poisoning. In patients with Itai-itai disease, urinary trehalase activity correlates inversely with tubular resorption of phosphorus (Nakano et al., 1987) and there is a statistical correlation between urinary trehalase and other urinary indicators of renal tubular dysfunction, such as glucose, $\beta_2$-microglobulin, cadmium, and $\alpha$-amino nitrogen, in inhabitants of chronic cadmium-polluted areas in Japan (Nogawa et al., 1980).

Other renal effects in cadmium workers include glucosuria, aminoaciduria, impaired concentrating abilities, and hypercalciuria (Kazantzis et al., 1963; Scott et al., 1976), which may cause disturbances in bone and calcium metabolism (section 8.2.2). The hypercalciuria leads to renal stone formation in some workers (section 8.2.2.1). An increased excretion of amino acids, particularly of serine and threonine, has been shown in industrial workers (Clarkson & Kench, 1956), but the amino acid excretion pattern was not consistent. In a cadmium worker with osteomalacia (Kazantzis, 1979), there was an increase excretion of hydroxyproline, which could be an effect of changes in collagen metabolism related to the bone disorder (section 7.2.4).

The renal effects of cadmium that lead to proteinuria may progress and, in some cases, with high exposures, lead to an increase in blood creatinine. This has contributed to a higher-than-expected mortality rate among highly exposed workers (Kjellström et al., 1979). In a Swedish battery factory, there were four deaths from nephritis or nephrosis among 185 workers, all of whom had been exposed to cadmium for more than 15 years. The expected number of deaths due to these causes in this group of workers was 0.4 ($P = 0.05$) (Andersson et al., 1984). In a study of about 6995 cadmium-exposed British workers (Armstrong & Kazantzis, 1983), there were 10 deaths due to nephritis or nephrosis, whereas 15.3 deaths were expected. In the subgroup with the highest exposure (211 workers), one death occurred (0.3 expected). A 5-year follow-up of this study (Kazantzis et al., 1988) confirmed no excess mortality from nephritis and nephrosis (ICD 580-584), the number of observed deaths now having increased to 16 (18.9 expected). Armstrong & Kazantzis (1985) conducted a case control study of this cohort in which a more detailed assessment of the past exposure of workers was obtained. There was a marginally increased, but not statistically significant, risk from nephritis and nephrosis (ICD 580-584) in workers with "ever high" or "ever medium" exposure to cadmium.

Existing studies of mortality from nephritis/nephrosis have been based upon epidemiological studies of renal failure given as the underlying cause of death on death certificates. With the advent of kidney dialysis and transplantation, patients with kidney failure frequently survive and die of other causes. If kidney failure is indicated at all on their death certificates, it is frequently given as a contributing rather than underlying cause of death.

No studies on the contribution of cadmium-induced renal effects to morbidity, absence from work, etc., have been published, although one study (Vorobjeva & Eremeeva, 1980) reported increased cardiovascular disease and related increased work absence among cadmium workers (section 8.2.4).

#### 8.2.1.2 *In the general environment*

In Japanese cadmium-polluted areas, signs of renal dysfunction very similar to those in cadmium-exposed industrial workers have been found. Proteinuria and glucosuria were found to be common (30-80%) among the exposed people in one area (Ishizaki, 1969) and less common among people living in control areas and areas bordering a polluted area. In the exposed groups, a positive correlation between age and the prevalence of signs was also seen (section 8.3.2). Due to the cumulative nature of cadmium, the total dose is directly correlated to age.

With the large number of elderly people and women included in the groups exposed in the general environment, factors other than cadmium that can affect renal function may make direct comparisons with industrial workers difficult. Nevertheless, the tubular proteinuria (Shiroishi et al., 1977), aminoaciduria, and other signs of renal tubular damage (Saito et al., 1977; Nogawa et al. 1984) were very similar to the findings for industrial workers. In addition, as in the case of exposed workers, elevated urinary excretion of metallothionein occurs as a result of environmental cadmium exposure (Tohyama et al., 1981b).

Among cadmium-exposed people in the general environment, the mean urine $\beta_2$-microglobulin excretion was highest (Shiroishi et al., 1977) in patients with Itai-itai disease (section 8.2.2.2). Many of the Itai-itai patients also have signs of decreased glomerular filtration, as indicated by decreased urea clearance (Nakagawa, 1960) and increased serum creatinine (Nogawa et al., 1979).

#### 8.2.1.3 Methods for detection of tubular proteinuria

Determination of total protein and electrophoretic analysis of concentrated urinary protein were originally the common methods for the detection and diagnosis of tubular proteinuria. Quantitative immunological methods (detection limit, 0.002-0.003 mg/litre) for the measurement of $\beta_2$-microglobulin (Evrin et al., 1971) and RBP (Bernard et al., 1982b) in urine ("normal" level about 0.1 mg/litre) are available, and these methods facilitate the detection of tubular dysfunction. An electrophoretic method with reasonable sensitivity (0.8 mg/litre) for measuring specific proteins in the urine utilizes staining of proteins in sodium dodecyl sulfate acrylamide gel electrophoresis (Nomiyama et al., 1982b). More recently, radio-, latex-, and enzyme-linked immunoassays have been developed (Evrin et al., 1971; Bernard et al., 1982a; Carlier et al., 1981). However, the disadvantage with $\beta_2$-microglobulin as a marker for renal tubular dysfunction is that this protein is unstable if urinary pH is less than 5.5.

RBP has an advantage over $\beta_2$-microglobulin, particularly for screening purposes, in that serum levels and, hence, excretion are not affected as readily by concomitant immunological disease and are more stable at an acidic pH (Bernard et al., 1982b). An enzyme-linked immunosorbent assay (ELISA) for urinary RBP has also been described (Topping et al., 1986).

Common tests for qualitative determination of proteinuria, such as paper tests (dip sticks), the nitric acid test, and the boiling test, should not be used for screening cadmium-induced proteinuria, since positive readings will be obtained only at fairly high urine protein concentrations (Piscator, 1962). Trichloroacetic acid (TCA) or sulfosalicylic acid (SA) can be used for qualitative tests, but a negative result does not exclude a moderate increase in low molecular weight proteinuria.

#### 8.2.1.4 Significance of cadmium-induced proteinuria

More than 70% of proteins with a molecular weight less than 15 000 but less than 5% of those with a molecular weight greater than 40 000 pass through the glomerular membrane (Squire et al., 1962). The glomerular filtrate contains relatively large amounts of plasma proteins, which are normally almost completely reabsorbed in the proximal tubules and only small amounts are found in the urine. The increased excretion of tubular proteins in cadmium nephropathy is thought to be due mainly to a decreased tubular reabsorption capacity. This provides early evidence of renal tubular dysfunction.

The concentration of albumin in normal serum is about 25 000 times higher than the concentration of $\beta_2$-microglobulin (Table 15). In spite of the fact that very little of the albumin but about 80% of the $\beta_2$-microglobulin is filtered through the glomeruli (Maack et al., 1979), the filtered amount of albumin is still higher than the amount of $\beta_2$-microglobulin (Table 15).

Only a small fraction of the albumin or the other proteins is excreted in normal urine due to the normally efficient (more than 99%) tubular reabsorption of all proteins (Table 15). The urinary albumin excretion is about 100 times greater than the $\beta_2$-microglobulin excretion (Table 15).

$\beta_2$-Microglobulin is a subunit of a major immunoglobulin complex with a molecular weight of 12 000 and normally occurs in serum at a concentration of approximately 2.0 mg/litre. In the case of a decreased glomerular filtration rate, the serum concentration of $\beta_2$-microglobulin will also increase. In certain conditions, for example, where excessive production occurs as in some cancers and autoimmune disorders, serum levels increase, the tubular capacity for reabsorption may be exceeded, and the concentration in the urine will rise. In tubular dysfunction, the capacity for absorption is impaired and, again, this is reflected by an increased excretion in the urine. Such renal disorders include the congenital and acquired Fanconi syndrome, diabetic nephropathy, certain cases of reflux nephropathy, and advanced glomerular disease (Squire et al., 1962). A raised urinary excretion of $\beta_2$-microglobulin or other low molecular weight protein is not, therefore, specific to renal dysfunction induced by cadmium, and a differential diagnosis should be considered in all cases where this occurs.

The "normal" average urinary excretion of $\beta_2$-microglobulin measured in several populations was in the range 0.05-0.1 mg/24 h (or mg/litre or mg/g creatine) (Kjellström & Piscator, 1977). Below age 65, there was very little or no change with age in the urinary $\beta_2$-microglobulin (Kjellström & Piscator, 1977), a fact confirmed by later studies (Tsuchiya et al., 1979; Kowal & Kraemer, 1982). In all of the studies, some high individual values were found in the age group above 65 years, and the studies of Tsuchiya et al. (1979) and Kowal & Kraemer (1982) reported age-regression coefficients indicating an increase with age. However, there was wide variation with age and the average urinary $\beta_2$-microglobulin levels in the oldest age groups (above 65 years) were only 10-20% higher than in the other age groups (Kowal & Kraemer, 1982). The prevalence of increased low molecular weight proteinuria (above 0.5 mg/litre) was less than 5% in these studies, but in a control group of people above age 80 from a Japanese cadmium-polluted area (section 8.3.2.2) the prevalence was about 15%.

#### 8.2.1.5 *Glomerular effects*

Although renal tubular dysfunction with its accompanying low molecular weight proteinuria is thought to be the most prominent renal effect of cadmium, the $\beta_2$-microglobulinuria is sometimes accompanied by the excretion of high molecular weight proteins such as albumin (molecular weight, 69 000). This albuminuria may occasionally occur as a result of cadmium exposure without any concomitant increase in the urinary excretion of low molecular weight proteins (Bernard et al., 1976, 1979); this indicates that cadmium, in some cases, may produce a change in the glomerular permeability to larger proteins.

Cadmium may also affect the glomerular filtration rate (GFR). Friberg (1950) reported decreased inulin clearance in cadmium-exposed battery workers. Elinder et al. (1985a) measured GFR by chromium-EDTA in 17 workers previously exposed to cadmium fumes. They found a significant negative correlation between decreasing GFR and tubular reabsorption loss, and reported that GFR decreased with increasing cumulative exposure to cadmium fumes. The urinary clearance of $\beta_2$-microglobulin increased with decreasing GFR.

Several other occupational studies have reported increased serum concentrations of creatinine and/or $\beta_2$-microglobulin, indicating reduced GFR, in cadmium-exposed workers (Thun et al., 1989; Roels et al., 1989).

Thun et al. (1989) found a small increase in mean serum creatinine in a group of 45 cadmium-exposed workmen. Serum creatinine also increased with cadmium dose, suggesting decreased glomerular function. Cadmium dose remained the important predictor of serum creatinine even after controlling for age, blood pressure, body size, and other extraneous factors.

Roels et al. (1989) measured the serum creatinine and serum $\beta_2$-microglobulin levels of 23 workers, removed from cadmium exposure, on several occasions over a period of six years. The average yearly decrease in GFR was estimated to be 6 ml/min per 1.73 $m^2$, which is considerably more than the normal value (< 1 ml/min per 1.73 $m^2$) and significantly more than that of a control group examined at the same time.

There is also evidence of glomerular effects in people exposed to cadmium in the environment. Nogawa et al. (1980) suggest that a reduction in creatinine clearance may be detected at the early stage of cadmium poisoning in a polluted area. In addition, Nogawa et al. (1984) reported a significant correlation between decreased tubular reabsorption of phosphate and decreased GFR in farmers living in a cadmium-polluted area.

The mechanism for the glomerular effects from cadmium is uncertain. It has been suggested that cadmium-induced tubular damage leads to a certain degree of interstitial nephritis which in turn results in a decreased GFR (Elinder et al., 1985a). It has also been proposed that cadmium exerts a direct toxic effect on the glomerulus (Roels et al., 1989).

#### 8.2.1.6 *Relationship between renal cadmium levels and the occurrence of effects*

The number of reports of renal pathology in autopsy cases and renal biopsies that contain data on kidney cortex concentrations of cadmium is small. Thus, it is difficult to establish a dose-response relationship between cadmium content and pathology or dysfunction.

Nomiyama (1977) summarized data from 26 cadmium-exposed workers and 16 cadmium-exposed people from the general environment. The criteria for choosing the subjects were that they possessed high renal and/or high liver concentrations of cadmium, morphological studies on the kidney had been performed, and data were available on the occurrence of proteinuria while the person was alive. Among the 42 cases reviewed, those exhibiting slight or no proteinuria and no morphological alterations had higher concentrations of cadmium in the renal cortex than non-exposed people. Most cases with morphological changes plus proteinuria had lower renal cadmium concentrations that those without proteinuria and/or morphological changes. In more recent studies, Ellis et al. (1985) found that in cases of renal dysfunction the mean liver and kidney cadmium values for retired workers were lower than those for active workers. These findings are similar to those from animal studies (section 6.5.1.2), where kidney concentrations levelled off or even declined in the presence of kidney damage.

The use of *in vivo* neutron activation analysis has facilitated the study of the relationship between renal cadmium levels and occurrence of effects (Ellis et al., 1981a; Roels et al., 1981b). However, the data must be assessed with caution as the accuracy of this method has not yet been fully determined. For instance, the exact location of the kidney needs to be known. Erroneously low renal cadmium levels were reported by Roels et al. (1981b) due to an error in adjusting for the distance between skin and kidney (Roels et al., 1983a). Data from Ellis et al. (1981a) and Roels et al. (1983a,b) have shown that few cases with increased urinary $\beta_2$-microglobulin concentrations are seen when the level of renal cortex cadmium is less than 150 mg/kg tissue and that of liver cadmium is less than 40 mg/kg. There is a pattern of liver and kidney cadmium levels increasing simultaneously until the average renal cortex cadmium concentration is about 300 mg/kg and the average liver level is about 60 mg/kg. At higher liver levels, the renal cortex level is disproportionately low in most cases, and, in addition, many of these workers have increased urinary $\beta_2$-microglobulin.

Skerfving et al. (1987) measured kidney cortex cadmium levels by X-ray fluorescence in a group of 20 workers from a factory producing alkaline batteries and found an average value of 147 mg/kg (range 53-317). When compared to a control group, these workers had higher average urine levels of cadmium (5.4 vs 0.8 nmol/mmol creatinine) and $\beta_2$-microglobulin (14.6 vs 6.6 $\mu$g/mmol creatinine). Six workers had $\beta_2$-microglobulin levels exceeding 22 $\mu$g/mmol creatinine. Due to selection procedures the results are, however, not predictive for cadmium-exposed workers in general. Nevertheless, it is clear that there are no significant correlations between levels of cadmium or $\beta_2$-microglobulin in urine and cadmium levels in the kidney. These results suggest that there is a relationship between renal cadmium and occurrences of effects on a group basis but renal cadmium levels *per se* are not always predictive of pathological effects on an individual level.

### 8.2.1.7 Reversibility of renal effects

The potential for reversibility of renal effects has been studied in populations of workers with occupationally induced cadmium nephropathy as well as in residents of cadmium-polluted areas.

#### a) Occupational exposures

In a group of 40 workers heavily exposed to cadmium, it was found, using qualitative methods, that proteinuria was persistent and sometimes even increased several years after cessation of exposure (Friberg & Nystrom, 1952; Piscator, 1966a).

Tsuchiya (1976) studied a group of 13 workers who had been exposed to cadmium fumes (133 $\mu g/m^3$) and who had proteinuria (determined by the trichloroacetic acid method) and abnormal electrophoretic urine patterns ($\beta_2$-microglobulin levels above 40 000 $\mu g$/litre). A 10-year follow-up study of five of these patients was carried out after improvements had been made in their working environment (cadmium fumes, 20 $\mu g/m^3$). The proteinuria was reversed (measured using a single radial immuno-diffusion method) in three of the five patients: $\beta_2$-microglobulin values were 3500, 2600 $\mu g$/litre, and not detectable (limit of detection 2000 $\mu g$/litre), and retinol-binding proteins (RBP) were not detectable (limit of detection 500 $\mu g$/litre). In addition, there were improvements in the remaining two patients ($\beta_2$-microglobulin, 9700 and 5500 $\mu g$/litre; RBP, 34 000 and 120 000 $\mu g$/litre). Tsuchiya (1976) suggested that the difference in the period of exposure to cadmium was the reason for this difference in the degree of recovery from the effects of cadmium.

Stewart & Hughes (1981) reported on similar cases from a British pigment factory. Despite the fact that exposure ceased many years earlier, grossly elevated $\beta_2$-microglobulin levels were still detected.

Using quantitative methods to detect proteinuria, Roels et al. (1982) found that total proteinuria in 19 workers was unchanged 4 years after exposure had ceased. In those 11 workers for whom urinary $\beta_2$-microglobulin was measured before and after cessation of exposure, levels were invariably increased. Eight of the workers had abnormal levels of $\beta_2$-microglobulin even before cessation of exposure, whereas three had normal levels before cessation and developed abnormal levels afterwards.

Roels et al. (1989) examined 23 workers once a year for 5 years after removal from exposure to cadmium. These workers had been exposed to cadmium for periods of 6 to 41.7 years (mean 25 years), and their first follow-up examination took place when they had been removed from exposure for an average of 6 years. Their mean age at that time was 58.6 years (range 45.5-68.1 years). Cadmiumuria in these workers had been assessed three years previously by measuring the cadmium levels in liver and kidney using neutron activation analysis. The cadmium concentrations (mg/kg wet weight) in the liver and kidney cortex ranged from 24 to 158 (mean 61) and from 133 to 355 (mean 231), respectively. Although cadmium concentrations in the blood and urine decreased significantly over the five-year period, urine concentrations of albumin, $\beta_2$-microglobulin, and RBP did not change significantly.

Harada (1987) conducted studies on the health status of seven workers exposed prior to 1972 to high cadmium levels in a cadmium sulfide dye manufacturing factory. These workers were examine for 15 years, improvements having been made to working conditions in 1974 which led to markedly decreased cadmium exposures. The cadmium content of the blood and urine declined after the improvements in working conditions but increased again when production rose. The working condition improvements resulted in a marked reduction in urine $\beta_2$-microglobulin level in five workers (e.g., from 1272 to 520 $\mu$g/g creatinine in 1 year in one worker and from 2090 to 503 $\mu$g/g creatinine in 6 months in another), but there was elevated urinary $\beta_2$-microglobulin excretion when production increased. One worker had fairly constant near-normal urinary levels of $\beta_2$-microglobulins (55 to 183 $\mu$g/g creatinine) regardless of workplace improvements or production levels. Initially, four workers had low GFR values, but none of the seven workers showed any decrease in glomerular filtration during the 15-year follow-up period. TRP rates decreased in three workers but remained relatively unchanged in the other four. These changes in GFR and TRP seemed to be independent of cadmium exposure levels.

Elinder et al. (1985b) found that urine cadmium excretion decreased in 14 out of 19 workers re-examined at least once five years or more after exposure to cadmium, but renal tubular function, as measured by urinary $\beta_2$-microglobulin excretion, had deteriorated or not improved in nearly all of the workers.

Thun et al. (1989) concluded that "time since last exposure to cadmium" was not an important determinant of renal outcome whether considered on its own or together with the cadmium dose. In their study of 45 workers at a plant that recovered cadmium from industrial waste, 9 out of 15 workers with the highest $\beta_2$-microglobulin excretion had not been exposed to cadmium for at least five years and one for 45 years. This study suggested that if cadmium nephropathy is reversible, the recovery is so slow as to be indiscernible after decades of non-exposure.

Ellis et al. (1985) showed that the liver cadmium levels in workers no longer exposed to cadmium gradually declines. Persistence of renal tubular dysfunction after cessation of exposure may reflect the level of body burden and the transfer of cadmium from liver to kidney.

*b) Residents of cadmium-polluted areas*

Reversibility of renal tubular dysfunction has been investigated in residents of cadmium-polluted areas in Japan. Kasuya et al. (1986) carried out a comparative study of urine $\beta_2$-microglobulin determinations made in 1975 and 1985 for 93 people with Itai-itai disease and their family members. Urine $\beta_2$-microglobulin levels improved in the group with $\beta_2$-microglobulin levels of 1000 $\mu$g/g creatinine or less but worsened in the group with $\beta_2$-microglobulin levels of 3000 $\mu$g/g creatinine or more. Most of the people who recovered were aged 39 years or less and had been resident for 30 years or less. It was considered that mild renal dysfunction among young individuals was reversible.

Saito (1987) measured the urine $\beta_2$-microglobulin levels of residents of cadmium-polluted areas for 3 years after improvements had been made in the level of soil contamination and compared the results with determinations obtained 7 years previously. During this 3-year period, the urine $\beta_2$-microglobulin levels tended to remain unchanged in people with a concentration of around 1000 $\mu$g/litre.

The reversibility of $\beta_2$-microglobulinuria, glucosuria, and aminoaciduria was examined in 74 inhabitants (32 males and 42 females) over 50 years of age who lived in a cadmium-polluted area. Examinations were conducted just after the cessation of cadmium exposure and 5 years later. The geometric mean concentrations of $\beta_2$-microglobulinuria, glucosuria, and aminoaciduria indicated a significant increase in excretion during the 5-year period. In cases where the level of $\beta_2$-microglobulinuria exceeded 1000 $\mu$g/g creatinine at the time cadmium exposure ceased, evidence was found indicating significant increases in proteinuria after 5 years, whereas in cases where the excretion of $\beta_2$-microglobulin had been less than 1000 $\mu$g/g creatinine no significant changes were observed (Kido et al. 1988).

### 8.2.2 Disorders of calcium metabolism and bone effects

#### 8.2.2.1 In industry

Friberg (1950) observed 7 cases of renal stones among 43 cadmium workers and drew attention to the possibility of renal stones being associated with exposure to cadmium. Ahlmark et al. (1961) found that 44% of a group of 39 cadmium workers exposed to cadmium oxide dust for more than 15 years had a history of renal stone formation. Nine stones from six workers were analysed, and in four workers the stones were composed of basic calcium phosphate (Axelsson, 1963). There was an increase in the mean calcium excretion rate in the cadmium-exposed group as compared to a control group. Kidney stones were also found in 12 out of 43 British workers at an accumulator factory (Adams et al., 1969). It is noteworthy that in both of these studies (Axelsson, 1963 and Adams et al., 1969) there was a higher prevalence of renal stones in workers without proteinuria than in those with proteinuria. The men with proteinuria had, as a group, increased urinary excretion of calcium and phosphate, whereas in the group without proteinuria there were a few cases with hypercalciuria. Hypercalciuria (81%) and renal stones (19%) were also reported among 27 coppersmiths exposed to cadmium (Scott et al., 1976, 1978).

Elinder et al. (1985a) found an increased prevalence of renal stones among cadmium workers with tubular proteinuria, and Mason et al. (1988) observed decreased renal reabsorption of calcium among cadmium alloy workers.

Seven of the 12 workers in a cadmium pigment factory investigated by Kazantzis et al. (1963) were found to have a urinary calcium excretion rate greater than 300 mg/day (at least in 1 of 2 specimens). There was no evidence of excessive calcium intake in these men. Five of these seven workers had been exposed to cadmium compounds for more than 25 years and also had tubular proteinuria. The remaining two, who had been exposed for 2 and 12 years, respectively, had no other abnormality except for a urinary calcium excretion of 308 and 403 mg/day. Follow-up was possible with six of the twelve men, including all five with hypercalciuria and proteinuria (Kazantzis, 1979). Six of the seven who had hypercalciuria when first examined continued to have a raised urinary calcium excretion, and one further worker developed hypercalciuria during the follow-up period. All those with hypercalciuria also had tubular proteinuria, although this was marginal in one of the workers. Blood calcium levels remained within normal limits in all cases (Kazantzis, 1979). The occurrence of disordered calcium metabolism in all seven men followed-up for a number of years makes it very likely that a common environmental factor, such as occupational exposure to cadmium compounds, was causative.

The data of Kazantzis (1979) agree with the findings of hypercalciuria among 27 coppersmiths with high cadmium exposure (Scott et al., 1976, 1978). Scott et al. (1980) reported that in 15 cadmium-exposed men the amount of calcium in the whole body was lower than that of controls and decreased with duration of an increased exposure to cadmium. The cadmium-induced hypercalciuria could be reduced by thiazide treatment (Scott et al., 1979).

In a study by Thun et al. (1989) of workers at a plant that recovered cadmium from industrial waste, 8 of 45 exposed workers had experienced kidney stones, in contrast to one of 32 unexposed workers. Increase in the urinary excretion of $\beta_2$-microglobulin and RBP was accompanied by decreased renal tubular reabsorption of calcium and phosphates.

In contrast to the above findings, a low urinary calcium excretion was detected in 47 out of 81 workers with exposure to a variety of cadmium compounds and also cadmium oxide fumes (Tarasenko & Vorobjeva, 1973; Tarasenko et al., 1975). The 24-h excretion of calcium in these workers was below 100 mg, compared with 115-210 mg in a control group of 21 people. Blood calcium values were within the normal range in all cases.

Radiological examination was performed on 32 workers exposed for 4-20 years to cadmium compounds (concentrations ranging from 0.1 to 5.5 mg/m$^3$). All of them complained of pains in the bones. Pseudofractures suggestive of osteomalacia were seen in two workers exposed for 16 and 19 years, but no histological confirmation of osteomalacia was obtained. Radiological appearances described as enostosis were reported in five cases and periosteal proliferation and consolidation in a further three cases (Tarasenko & Vorobjeva, 1973; Tarasenko et al., 1975).

Horstowa et al. (1966) performed radiological examination of the skeleton in 26 alkaline battery workers with signs of chronic cadmium intoxication out of 80 workers exposed to 0.13-1.17 mg cadmium/$m^3$ for 1-12 years. Seven of these workers had proteinuria detected by sulfosalicyclic acid; pseudofractures were found in 3 workers, sclerotic foci in 13, and osteoporosis in 10. In another alkaline battery factory, where a number of cases of severe cadmium poisoning were diagnosed (Friberg, 1950), X-ray examinations revealed no signs of bone disease.

One of the workers with multiple tubular defects studied by Kazantzis et al. (1963) developed osteomalacia confirmed by histological examination 10 years after the initial investigation (Kazantzis, 1979). He previously displayed hypercalciuria but, at the time of diagnosis, his urinary calcium excretion was low. Extensive investigation failed to reveal any of the other generally accepted causes of osteomalacia such as malabsorption or nutritional deficiency.

In a study of 43 workers at a battery plant, one worker developed osteomalacia without evidence of malabsorption or nutritional deficiency but with multiple renal tubular defects (Adams et al., 1969). Another case of osteomalacia from the same factory was subsequently detected in a man who had been a cadmium battery worker for 40 years (Adams, 1980). Eight years before retirement he had a partial gastrectomy due to a duodenal ulcer. Proteinuria was first diagnosed 6 years before retirement, but otherwise he was in "apparent good health" and "on a balanced diet" until 8 years after retirement. He was then frail, had pains in his legs, and a "waddling gait". His serum alkaline phosphatase level was increased, X-rays showed generalized osteoporosis, and a bone biopsy showed osteomalacia. After 1 year of treatment with large doses of vitamin D, he could walk well again.

It also seems likely that the six workers exposed to cadmium oxide dust described by Nicaud et al. (1942), who had pains in the back and limbs and showed multiple pseudofractures on radiological examination, suffered from osteomalacia. More detailed data on these workers was presented by Valetas (1946), who also pointed out that "massive doses" of vitamin D were needed to improve the symptoms. It took several months for improvement to occur and the vitamin D treatment had to be maintained for several years to keep the workers in stable health. Valetas (1946) concluded that this bone disease was caused by occupational exposure to cadmium. Eight workers with 8-30 years of exposure to lead dust and cadmium oxide fume and dust (Gervais & Delpech, 1963) were also found to have multiple pseudofractures and pains in the back, thorax, and legs. Very limited biochemical investigations were carried out, but in four cases proteinuria was found. The authors suspected that lead exposure led to the observed effects.

#### 8.2.2.2 *In the general environment*

Bone disease and abnormalities of calcium metabolism from exposure to cadmium in the general environment have only been noted in people in Japan with the clinical syndrome referred to as Itai-itai disease. The main characteristics of the disease are osteomalacia[1] and osteoporosis[2] with a tendency to fractures accompanied by severe pain and renal tubular dysfunction. The results of epidemiological and clinical investigations indicate an association with cadmium exposure, although the Co-operative Research Committee on Itai-itai Disease (1967) stated that "malnutrition (low protein, low calcium diets) and multiple pregnancies may also be involved".

Itai-itai disease is an endemic bone disease prevalent in the basin of the Jinzu river, which runs through the central part of Toyama Prefecture in West-Central Japan (Kono et el., 1956). It is characterized by osteomalacia in combination with renal tubular dysfunction in most cases. Patients also have osteoporosis and one of the most characteristic symptoms is severe bone pain. Hagino & Yoshioka (1961) reported that high concentrations of cadmium, lead, and zinc were present in autopsy tissues from people with Itai-itai disease and in the everyday foods of the endemic area.

---

[1] Osteomalacia is characterized by inadequate mineralization of bone matrix, resulting in an increase in the relative amount of osteoid tissue. It represents the adult counterpart of childhood rickets (Robbins et al. 1984).

[2] Osteoporosis is defined as an excessive but proportional reduction in the amounts of both the mineral and matrix phases of bone unaccompanied by any abnormality in structure of the residual bone (Robbins et al., 1984).

Systematic epidemiological investigations, which included extensive mass health examinations as well as case control studies on both patients and controls, started in 1962 (Cooperative Research Committee on Itai-itai Disease, 1967). It was reported that Itai-itai disease in Toyama Prefecture was restricted to a limited area (Fuchu area) irrigated by the Jinzu river, the geographical distribution of the patients being consistent with the levels of cadmium concentration in the paddy fields, and that the concentrations of cadmium in urine were higher in patients than in controls. The total number of patients was estimated in 1955 by the Toyama Prefecture to be 41 out of a total of 1666 residents (849 women) (Cooperative Research Committee on Itai-itai Disease, 1967). The major source of cadmium pollution in the area was a mine 50 km upstream from the endemic area (Japan Public Health Association, 1968).

The age and sex distribution of the patients displayed a very distinct pattern. Clinically apparent cases were limited to women over 40 years of age who had given birth to many children (6 on average) and had lived in the area for more than 30 years. No detailed data on past patients were available, but it was estimated that the age of onset of the disease was probably between 35 and 65 years, and that almost 100 deaths had been reported up to the end of 1966. The incidence was presumably very high from 1936 to 1950 and at its highest in 1946 and 1947, but decreased thereafter even though the same cadmium exposure levels had been maintained. By March 1989, 150 cases of Itai-itai disease had been officially recognized as pollution-related disease. Whereas all cases have been reported in the Fuchu area, there have been a few suspected cases in 2 out of 12 cadmium-contaminated areas of Japan other than the Fuchu area (Table 7). Clinical features of five suspected cases from the Ikuno area matched those of Itai-itai disease and urine cadmium levels were very high (Nogawa et al., 1975). In one of these cases an autopsy was performed; the liver cadmium level was very high (75 mg/kg) but the renal cortex cadmium level was low (53 mg/kg) (Nogawa et al., 1975).

Takebayashi (1980a,b, 1983a,b, 1984) and Takebayashi et al. (1985, 1987a,b,c, 1988a,b,c,d) reported pathological findings in kidney and bone from autopsies of eleven elderly men and women (3 males and 8 females; 72-95 years of age) from Tsushima Island. The average levels of cadmium in the liver and kidney cortex were 92.4 mg/kg and 44.0 mg/kg, respectively. The authors considered the histological osteomalacia and renal tubulopathy noted in eight cases (Takebayashi, 1980, 1983a, 1984, Takebayashi et al., 1985, 1987a,b, 1988c,d) to be similar to Itai-itai disease from Toyama Prefecture.

However the Japan Cadmium Research Committee (1989), supported by the Japanese Environment Agency, concluded, after these eight cases had been examined by the expert group, that it was clinically difficult to diagnose them as osteomalacia.

According to the Japan Cadmium Research Committee (1989), diagnosed cases of Itai-itai disease were reported only in the Fuchu area of Japan. It denied the presence of osteomalacia in five cases in the Ikuno area, and stated that osteomalacia had not been observed "clinically" in Tsushima Island.

A study by Kido et al. (1989) indicates that exposure to cadmium could cause osteopenia, particularly in women. Bone density was measured in 28 women with Itai-itai disease, 92 men and 114 women with cadmium-induced renal dysfunctions, and 44 men and 66 women living in three different non-polluted areas using a microdensitometer. The values of indices corresponding to both cortical width and bone mineral content were significantly lower in Itai-itai disease patients than in cadmium-exposed women with renal dysfunctions or in non-exposed subjects. The cadmium-exposed women also showed a decrease in bone density compared with the non-exposed subjects. A significant decrease in bone density was also observed in cadmium-exposed men compared with non-exposed subjects, although the difference was not as clear as it was in women.

Reviews (in English) of Itai-itai disease have been produced by Tsuchiya (1969), Friberg et al. (1974), Tsuchiya (1978), and Nogawa (1981).

#### 8.2.2.3 *Mechanism of cadmium-induced bone effects*

The available data show that cadmium can affect calcium, phosphorous, and bone metabolism in both industrial workers and people exposed in the general environment. These effects may be secondary to the cadmium effects on the kidneys but there have been few studies of calcium metabolism in people with excess exposure to cadmium. The increased prevalence of renal stones reported from certain industries is probably one manifestation of the cadmium-induced kidney effects. It is not known if factors other than cadmium play a role.

Nogawa et al. (1987) reported that serum 1,25-dihydroxy-vitamin D levels were lower in Itai-itai disease patients and cadmium-exposed subjects with renal damage than in non-exposed subjects. The reduction in these levels was closely related to serum concentrations of parathyroid hormone and $\beta_2$-micro-globulin and to the percentage tubular reabsorption of phosphate (% TRP), suggesting that cadmium-induced bone effects were mainly due to a disturbance in vitamin D and parathyroid hormone metabolism.

Osteomalacia has been reported in a few heavily exposed industrial workers and people with Itai-itai disease. The industrial cases are mainly male, whereas Itai-itai patients are almost exclusively female. However, the clinical features and biochemical findings are similar, except that Itai-itai patients may also suffer from ostoporosis.

A possible mechanism for the development of osteomalacia has been proposed (Kjellström, 1986). It is known that normal calcium absorption in the intestines and normal bone mineralization is dependent upon 1,25-dihydroxycholecalciferol. Vitamin $D_3$ taken into the body is converted to 25-hydroxy-vitamin $D_3$ in the liver, and then to 1,25-dihydroxy-vitamin $D_3$ in the mitochondria of renal proximal tubular cells, this being the biologically active species. Cadmium accumulates in the proximal tubular cells, depressing cellular functions, and this may result in reduced conversion of 25-hydroxy- to 1,25-dihydroxy-vitamin $D_3$. This is likely to lead to decreased calcium absorption and decreased mineralization of bone, which in turn may result in osteomalacia.

### 8.2.3 Respiratory system effects

Cadmium workers may develop chronic injury to the respiratory system, depending on the level and nature of exposure. The development of such effects is often quite slow, so that they are apparent only after several years of exposure. The rate of development and severity appear to be roughly proportional to the time and level of exposure.

#### 8.2.3.1 Upper respiratory system

Chronic inflammation of the nose, pharynx, and larynx have been reported by Vorobjeva (1958) and Horstowa et al. (1966).

Anosmia is a frequent symptom in cadmium workers after prolonged exposure. This has been reported by, for instance, Valetas (1946), Friberg (1950), Baader (1951), Vorobjeva (1958), Tarasenko & Vorobjeva (1973), and Apostolov (1979), but was not observed by Tsuchiya (1967) or Suzuki et al. (1965).

#### 8.2.3.2 *Lower respiratory system*

Chronic obstructive lung disease of varying degrees of severity is frequently seen in cadmium workers. Friberg (1950) reported dyspnoea, impaired lung function with increased residual volume, and reduced working capacity in a group of 43 cadmium workers. Similar studies, which included the use of pulmonary function measurements, by Bonnell (1955), Buxton (1956), Kazantzis et al. (1963), and Adams et al. (1969) all showed impairment of respiratory function in groups of workers with prolonged exposure. The symptoms and findings were more suggestive of emphysema than bronchitis in these cases; they were commonly diagnosed as emphysema but pathological confirmation of this was rare (Smith et al., 1960).

Tarasenko & Vorobjeva (1973) reported the presence of increased lung markings in the chest X-rays of 17 out of 72 cadmium workers, which were interpreted as being due to diffuse interstitial fibrosis. Similar lung changes were observed in 21 out of 26 workers studied by Horstowa et al. (1966).

The presence of chronic obstructive respiratory disease in cigarette smokers exposed to an additional harmful environmental agent presents difficulties in determining the contribution made by the latter. Studies on the chronic respiratory effects of cadmium in the past have not always been standardized for smoking. Lauwerys et al. (1974a) did take smoking habits into consideration by matching his cadmium-exposed and control groups for smoking habits. They reported the presence of impaired lung function in a group of cadmium workers exposed for over 20 years, but not in those with shorter exposure. The degree of lung impairment found was small.

The effects on the lung increases the mortality of cadmium workers with high exposures (Kjellström et al., 1979; Armstrong & Kazantzis, 1983). In the latter study, the mortality for diseases coded as bronchitis (ICD 490-491) was related to the intensity of exposure, the group with the highest exposure having a highly significant (almost 4-fold) excess risk (observed 13 expected 3.4).

A 5-year follow-up of this study (Kazantzis et al., 1988) confirmed the earlier finding, the marked excess mortality being related to both intensity and duration of exposure. The follow-up revealed an excess mortality from emphysema, but this was seen only in the low-exposure group.

### 8.2.4 Hypertension and cardiovascular disease

Despite the abundance of data showing that under certain exposure conditions cadmium induces hypertension in animals, there are very few results available from studies of cadmium-exposed workers. Friberg (1950) examined 43 workers with a mean period of exposure to cadmium oxide dust of 20 years (air concentration, 3-15 mg/m$^3$) and 15 workers with a mean exposure period of 2 years. The study included physical and roentgenological examinations of the heart, electrocardiographic examination at rest and after exercise, and measurement of blood pressure. No increased prevalence of cardiac disease or pathological electrocardiographic changes were found. The majority of subjects had completely normal blood pressure, but since Friberg did not examine blood pressure in the control group, it is not possible to draw definite conclusions.

Chest examination and blood pressure measurements have also been reported in other studies (Bonnell, 1955; Bonnell et al., 1959; Kazantzis et al., 1963; Holden, 1969), but in no cases were there findings of cardiac disease or hypertension due to cadmium exposure. Hammer et al. (1972) found no relationship between exposure to cadmium and blood pressure in superphosphate workers.

Vorobjeva & Eremeeva (1980) examined 72 female and 20 male workers at a battery factory exposed to cadmium oxide dust at concentrations ranging from 0.04 to 0.5 mg/m$^3$. Blood pressure was measured and electrocardiograms taken, but there was no control group. The authors reported increased prevalence of hypertension and absence from work due to hypertensive and ischaemic heart disease among the exposed workers compared to what was considered normal. Furthermore, several types of abnormalities in the electrocardiogram of the exposed workers were observed: 39% showed tachycardia, between 11 and 13% were regarded as normal, and 26% had changes in the "R" spike (compared to the normal 7-9%). Increased QRS period was observed in 45% of the workers compared to the normal values of 14-16%. The data presented in this report are especially interesting in view of the evidence in rats (section 7.2.6) that suggests myocardial effects from cadmium exposure. The results of the study are, however, presented in a very condensed form and it is therefore difficult to draw clear-cut conclusions.

In a retrospective study of 311 male workers in an alkaline battery factory it was found that hypertensive workers had a longer employment time than an age-matched control group from the same work environment (Engvall & Perk, 1985). Again it is difficult to draw conclusions from this study. In a study of cadmium-exposed workers in the United Kingdom (Kazantzis et al., 1988), mortality from hypertensive disease (ICD 400-404) over the total study period from 1943 to 1988 was elevated but not significantly (49 deaths occurred as opposed to 41.3 expected). There was no relationship with intensity of exposure. However, mortality from cerebrovascular disease (ICD 430-438) was significantly lower (178 deaths occurred as opposed to 230.3 expected). These findings do not suggest any association between cadmium exposure and the development of hypertension.

In contrast, Thun et al. (1989) found that mean systolic and diastolic blood pressures were higher in 45 cadmium workers (134 and 80 mmHg, respectively) than in 32 male controls (120 and 73 mmHg respectively). Blood pressure was measured systematically by a single examiner on the right arm of subjects who had been seated for at least 15 min. Systolic but not diastolic blood pressure was significantly associated with cadmium dose in multivariate analyses.

Schroeder (1965, 1967) observed that people in the general population dying from hypertensive and/or cardiovascular disease had somewhat higher cadmium concentrations in liver and kidney tissues than people dying from other causes. He suggested that cadmium could be a causative factor for these diseases. Unfortunately, smoking habits were not accounted for and it is likely that this was a confounding factor. The same problem exists with a number of subsequent studies on hypertension and cadmium in tissues, blood, and urine.

A correlation between average air cadmium levels in cities in the USA and mortality associated with hypertension and heart disease has been reported (Carroll, 1966; Hickey et al., 1967). Again, several confounding factors such as smoking habits, air pollutants other than cadmium, and other environmental factors make it difficult to draw conclusions concerning the effects of cadmium.

In a study by Staessen et al. (1984), the confounding variables age, sex, body weight, and cigarette smoking were considered in a multiple regression analysis of systolic and diastolic blood pressure and the urinary excretion of cadmium and $\beta_2$-microglobulin. Negative correlations between blood pressure and urinary cadmium or $\beta_2$-microglobulin were found in some groups. As there was a very strong age effect on both blood pressure and urinary cadmium, the meaning of the negative correlations is not clear. In any case, these data do not support cadmium exposure as a cause of hypertension.

Shigematsu et al. (1979) could find no evidence that blood pressure was higher in polluted areas (1611 people sampled) of Japan compared with control areas (1826 people). In a comparison of blood pressure by prefecture (13 570 in the cadmium-polluted areas and 7196 in the control areas), the prevalence of hypertension was found to be high in the polluted area of one of the eight prefectures investigated. However, in the other seven prefectures, the prevalence of hypertension tended to be lower in the polluted areas (Japan Cadmium Research Committee, 1989).

In a study on cadmium-polluted areas in Japan by Nogawa et al. (1981b), the cerebrovascular disease mortality rate among people who had had cadmium-induced proteinuria was twice as high as that of people in the same area without proteinuria. However, the difference was not statistically significant. The number of men in the cohort with proteinuria was 81 and the number without proteinuria was 1109. Another study comparing administrative units containing polluted areas with those without such areas (Shigematsu et al., 1981, 1982, 1983) found no difference in the cerebrovascular disease mortality rates.

Data on a total population of 333 000 from both cadmium-polluted and non-polluted areas were collected retrospectively for a period of 6-30 years, based on vital statistics or death certificates (Shigematsu et al., 1982). The mortalities from all causes, including cardiovascular disease such as cerebrovascular and hypertensive disease, in the general population in the cadmium-polluted areas were no higher than, or in some cases even lower than, those in the non-polluted areas.

A mortality study of Shipham residents and of a nearby control village was reported by Inskip et al. (1982). The study population consisted of 501 Shipham residents of whom 278 had died over a 40-year follow-up period. Overall mortality was low in both villages compared generally with England and Wales. There was a small but statistically significant excess mortality rate in Shipham from hypertensive and cerebrovascular disease. The highest ratio of all was for genito-urinary disease in Shipham men (but, with only eight observed deaths, the result was only significant at the 10% level). The Standardized Mortality Ratios (SMR) for nephritis and nephrosis in both sexes were also slightly elevated, but there were only two deaths for each sex from this cause. In men, the numbers of prostatic and lung cancer deaths were approximately equal to the expected numbers, and in neither case was the SMR in Shipham greatly different from that in the control village.

### 8.2.5 Cancer

#### 8.2.5.1 In industry

A number of epidemiological studies have been published. In order to facilitate the interpretation of published data on the relationship between cadmium exposure and cancer, the studies have been grouped according to the types of industrial plants in which they have been conducted. In some cases, more than one study has been conducted at the same plant.

*a) Nickel-cadmium battery plants*

In an early study, Potts (1965) found that three out of eight deaths in a small cohort of nickel-cadmium battery workers in the United Kingdom with at least 10 years of exposure to cadmium oxide dust were from carcinoma of the prostate. This study was extended by Kipling & Waterhouse (1967) to include 248 men with at least one year of exposure to cadmium oxide dust. Four deaths from carcinoma of the prostate, including the three cases previously reported by Potts (1965), were observed as opposed to an expected number of 0.58.

Sorahan & Waterhouse (1983) carried out a further investigation of the same plant using a cohort of 3025 employees who started work between 1923 and 1975 and had a minimum employment period of one month. The method of regression models in life tables was used to compare duration of exposed employment in those dying from relevant causes with that of matched survivors in the same year of follow-up. No new evidence of an association between occupational exposure to cadmium and cancer of the prostate was found. However, there was an excess mortality from cancer of the respiratory system significant at the 5% level (89 cases, SMR = 127). As in other studies, data on smoking habits were not available and confounding factors were present in the form of exposure to nickel hydroxide and welding fumes so that no firm conclusions about the pulmonary carcinogenicity of cadmium could be drawn from this study.

Sorahan & Waterhouse (1983) reported on the incidence of prostatic cancer in a subgroup of 458 workers employed for at least 1 year in a job involving high exposure to cadmium oxide dust. Eight cancers were observed compared to two expected (SMR 400, $P < 0.01$). However, exclusion of the four cases previously reported by Kipling & Waterhouse leaves a non-significant excess incidence ($P = 0.21$), from which the investigators concluded that if cadmium oxide is potentially carcinogenic current risks are likely to be small.

In the most recent update of the nickel-cadmium battery plant workers (Sorahan, 1987), the earlier findings were confirmed and there was some evidence of an association between risk of death from lung cancer and duration of employment in jobs with high or moderate exposure among workers first employed in the period 1923-1946. However, among workers first employed from 1947 to 1975 (the group with the higher SMR for lung cancer), there was no evidence of such an association. The authors concluded that the findings do not suggest these nickel-cadmium battery workers had experienced an elevated lung cancer risk as a consequence of exposure to cadmium oxide dust.

In Sweden, Kjellström (1979) investigated the incidence of cancer among 269 male nickel-cadmium battery workers. All workers had been heavily exposed (on average about 1 mg cadmium/$m^3$) for five years or more to cadmium dust or fume, and were alive in 1959. Fifteen workers were found to have cancer between 1959 and 1975. It was calculated from national incidence rates that 16.4 new cases would have occurred; only 2 were prostatic cancers while 1.2 were expected. In a re-examination of the same cohort, there were 8 deaths from lung cancer with a non-significantly raised SMR of 133. The SMRs increased progressively with increasing latent periods without reaching statistical significance (Elinder et al., 1985c).

*b) Copper-cadmium alloy plants*

Copper-cadmium alloy workers in the United Kingdom who had heavy past exposure to cadmium oxide fume on two sites, one urban the other rural, were studied by Holden (1980a). There was an increased lung cancer mortality at the urban site (8 observed versus 4.5 expected) and a significant deficit at the rural site (2 observed; 7.8 expected). Vicinity workers in the urban plant, where the mean cadmium concentration averaged no more than 60 $\mu g/m^3$), also experienced a significantly increased lung and prostatic cancer mortality (36 observed; 26.1 expected).

A case control study was performed (Kazantzis et al., 1989) in the same copper-cadmium alloy plants where workers had experienced heavy past exposure to cadmium oxide fume and dust, which had given rise to a number of deaths coded as chronic cadmium poisoning. Before and during the period 1939-1945, cadmium oxide fume levels had been estimated to be up to 4 $mg/m^3$. Cases and controls were selected from the cohort previously studied by Holden (1980a). Personal interviews conducted with a small number of long-term employees revealed that arsenical copper had been additionally produced by adding bags of arsenic trioxide to the molten copper and stirring manually; this resulted in the evolution of dense white clouds of arsenic fume. The case control study showed no evidence of an increased risk of lung cancer associated with past cadmium exposure but an approximately two-fold excess risk associated with arsenic exposure.

Kjellström (1979) also investigated a cadmium-copper alloy plant in Sweden where workers had been exposed to cadmium oxide fumes and included 94 workers employed in 1940 or who started work after that year. Four cases of prostatic cancer occurred as opposed to 2.7 expected.

*c) Cadmium recovery plant in the USA*

An increase in prostatic cancer incidence was also found by Lemen et al. (1976) in a study of 292 male smelter workers heavily exposed to cadmium oxide dust or fumes. Air cadmium concentrations in 1973 were up to 24 $mg/m^3$ but generally below 1 $mg/m^3$. There were four deaths from cancer of the prostate (1.15 expected). There were also 12 deaths from lung cancer (5.1 expected); the difference was statistically significant.

Thun et al. (1985) expanded the Lemen et al. (1976) cohort to include 602 workers who had been employed at this cadmium production plant between 1940 and 1969 for at least 6 months. Exposure was to cadmium in baghouse dust, a by-product of zinc smelting which was processed to produce cadmium metal and cadmium oxide. The plant functioned as an arsenic smelter up to the end of 1925, and small quantities of lead, arsenic, thallium and indium were subsequently produced at intervals. The vital status of the workers was determined in 1978. A dose-response relationship was observed between lung cancer mortality and cumulative exposure and was statistically significant for workers whose exposure exceeded 2920 $mg/m^3$.days. The SMR for this group was 280. The lung tumours were, as far as can be determined, mostly of bronchogenic origin. The authors accounted for smoking habits by obtaining questionnaires from survivors or next-of-kin in 50% of the cohort members and for arsenic exposure by measuring arsenic in certain parts of the plant.

*d) Cadmium processing plants in the United Kingdom*

Kazantzis & Armstrong (1982) and Armstrong & Kazantzis (1983) investigated a large cohort of workers in England at 17 plants with processes using cadmium. The cohort comprised 6995 cadmium-exposed male workers born before 1940, first exposed before 1970, and not included in any previous mortality study. Jobs were assessed for each relevant year involving high, medium or low exposure to cadmium on the basis of discussions with hygienists and employees with knowledge of past working conditions, taking into account environmental and biological monitoring data (e.g., cadmium urine data > 20 mg/litre in the high-exposure group. The periods at risk of the study population were classified on the basis of these categories and recorded job histories into three groups: (i) those workers continuously employed for more than one year in a job assessed as entailing high exposure - "ever high"; (ii) those workers continuously employed for more than one year in a job assessed as entailing medium exposure, but who were never for more than one year in a high-exposure job - "ever medium", and (iii) all others. Actual deaths were compared with expected numbers calculated from mortality rates for the population of England and Wales corrected for regional variation. The 8th revision of ICD codes was used and results were expressed as SMRs.

Only 3% of the workers (about 200) were assigned to the "ever high" category. The mean duration of exposure was 11 years and the mean interval from initial exposure to the end of the follow-up was 27 years. The SMR (all causes) for the entire population was 97. There were no prostatic cancer deaths in the "ever high" and "ever medium" exposure categories (0.4 and 2.5, respectively, expected), and the number of deaths (23) in the "always low" group was close to the expected value. There was a small, but not statistically significant, excess of lung cancer in all categories, but in those with more than 10 years exposure in the "always low" category this excess was significant at the 5% level (100 observed, SMR 126). Since there was no correlation between increase in lung cancer risk and intensity of exposure, the authors concluded that it was unlikely that the excess in the "always low" group was due to cadmium.

A 5-year update of this study (Kazantzis et al., 1988) confirmed no excess risk from prostatic cancer over the total study period from 1943 to 1984 and no cases of prostatic cancer in the medium- or high-exposure groups. The SMR was 99 as opposed to the value of 90 in the initial study. However, there was now a significant excess lung cancer mortality (277 observed deaths, 240.9 expected), giving a SMR of 115 (95% confidence interval, 101-129). This excess risk was related to intensity of exposure, there being 12 deaths in the small high-exposure group as opposed to 6.2 expected (SMR, 194; 95% CI, 100-339), 41 deaths in the medium-exposure group and 224 deaths in the low-exposure group (SMR 121 and 112, respectively; not significant). While there appeared to be evidence of a dose-response relationship, it was not statistically significant. The increased cancer risk mainly involved those employed before 1940, rising with length of employment and with length of follow-up.

Further studies have been conducted on workers at these 17 plants (Armstrong & Kazantzis, 1985; Ades & Kazantzis, 1988). A case control study on lung cancer was carried out on workers in a large lead-zinc-cadmium smelter. These workers formed 64% of the cohort of 6995 men, and the study included 70% of the lung cancer deaths observed in the cohort as a whole (Ades & Kazantzis, 1988). There was a significant excess lung cancer risk among the smelter workers, and a significant trend with increasing duration of employment, particularly evident among those employed for more than 20 years. Quantitative estimates of exposure to cadmium and ordinal rankings for lead, arsenic, zinc, sulfur dioxide and dust were used to calculate cumulative exposures from job histories. However, matched logistic regression analysis showed that the increasing risk of lung cancer associated with increasing length of employment could not be accounted for by cadmium exposure and did not appear to be restricted to any particular process or department.

*e) Summary of industrial studies*

Increased mortality from lung cancer has been observed in several occupational cohorts exposed to cadmium, and there is some evidence of dose-response relationships in two of the examined populations. Case control studies have not given support for such a relationship. It is difficult to reach a firm conclusion about causality, because in all of the occupational cohorts there has been simultaneous exposure to other potential carcinogens (e.g., nickel, arsenic, polyaromatic hydrocarbons) or other environmental pollutants (e.g., sulfur dioxide). Information on tobacco smoking is inadequate or entirely absent in all except two studies. Investigations of the relationships between cadmium exposure and prostatic cancer are inconclusive.

### 8.2.5.2 *In the general environment*

Elevated cadmium levels have been found in the liver and kidneys of patients with bronchogenic carcinoma (Morgan, 1970; Morgan et al., 1971). However, the authors stressed the possibility that differences observed could reflect the effect of smoking (section 5.1.3) or could represent a non specific association.

A study of the causes of death in areas of high cadmium exposures in Japan (Shigematsu et al., 1982) revealed no difference in age-adjusted cancer mortality rates between polluted and control areas of the same prefecture. The mortality rate due to prostatic cancer was elevated in two areas but only achieved statistical significance ($P < .01$) in one area. It was not significant in two areas including Toyama prefecture, which has the largest area of pollution.

## 8.2.6 Mutagenic effects in human cells

An increased frequency of chromosomal aberrations in somatic human cells is considered to be evidence of some exposure to mutagenic agents. Shiraishi (1975) noticed an increased frequency of chromosomal aberrations in lymphocytes obtained from 12 Itai-itai patients compared to 9 female controls. However, this observation was not confirmed by Bui et al. (1975) who examined cells from 4 Itai-itai patients and 4 controls.

Among cadmium workers, an increased prevalence of chromosomal aberrations, compared to controls, was reported by Deknudt & Leonard (1975) and by Bauchinger et al. (1976), whereas no such effect was seen by O'Riordan et al. (1978). In none of these occupational studies was the actual exposure to cadmium measured, and the possible confounding effect from other industrial chemicals and smoking was not considered.

Nogawa et al. (1986) did not find evidence for increased sister chromatid exchange in people exposed to cadmium in the general environment. IARC (1987a,b) reviewed the available evidence for mutagenic and related effects and noted the differences in results reported from different industrial environments.

In conclusion, it is not yet possible to say whether cadmium causes mutagenic effects in humans.

### 8.2.7 *Transplacental transport and fetal effects*

There have been few studies on the fetal toxicity of cadmium transported across the placenta. Maternal hypertension and decrease in birth weight have been associated with elevated levels of cadmium in the neonate (Huel et al., 1981). In addition, it is well-established that the babies of mothers who are cigarette smokers are smaller at birth than are those of non-smokers. The ratio of placental zinc to cadmium is positively related to infant birth weight in the case of pregnant smokers, and older pregnant smokers are at higher risk for impaired fetal growth than are younger ones (Cnattingius et al., 1985; Kuhnert et al., 1987a, Kuhnert et al., 1987b). Multiparity is related to an increased placental cadmium level in smokers and to a decreased placental zinc level in both smokers and non-smokers. These results have been interpreted as consistent with a depletion of zinc with increasing number of births and a progressive increase in cadmium in smokers because of the long half-life of cadmium (Kuhnert et al., 1988).

The cellular mechanisms and factors that influence transplacental transport of cadmium are not known. Metallothionein has been identified in the human placenta and in fetal membranes at term, and metallothionein synthesis is inducible in cultured trophoblasts by treatment with cadmium (Waalkes et al., 1984). This effect is seen with cadmium concentrations in the culture medium as low as 52.2 μg/litre (0.5 μmol/litre) (Lehman & Poisner, 1984). Higher levels of exposure to cadmium may have a direct toxic effect on the placenta. In a test system involving perfusion of maternal and fetal blood vessels in the isolated human placenta, it was shown that perfusion of the maternal circulation with cadmium at a concentration of 1.12 mg/litre (10 μmol/litre) resulted in the deposition of 2.5 μg cadmium per g placenta (22 nmoles/g), but very little of it was detectable in the fetal circulation. Perfusion of the maternal circulation with higher concentrations of cadmium produced placental cadmium concentrations of 11.2-16.8 μg/g (100-150 nmoles) with stromal oedema, syncytiotrophoblastic vesiculation and vacuolization of Hofbauer cells within 6-8 h, followed by placental necrosis. These changes were associated with a decrease in human chorionic gonadotropin release and decreased movement of zinc into the fetal circulation (Miller, 1986).

### 8.2.8 Other effects

Many other different symptoms and signs have been reported in humans exposed to cadmium. These include loss of appetite, loss of weight, fatigue, and increases in the erythrocyte sedimentation rate (ESR). Valetas (1946) reported details of the poisoning cases in a French accumulator factory, which were first described by Nicaud et al. (1942). In addition to the bone effects and the pains (section 8.2.2.1), Valetas mentioned that several workers experienced paraesthesia and involuntary muscular contractions. This could be an effect resulting from abnormal changes in the levels of serum electrolytes, such as calcium or potassium, which may in turn be caused by severe kidney damage.

Mild anaemia has been more frequently observed among cadmium-exposed workers than among controls (Friberg, 1950; Bonnell, 1955; Bernard et al., 1979).

More specific effects from cadmium are the yellow discoloration of the proximal part of the front teeth (Barthelemy & Moline, 1946; Valetas, 1946; Princi, 1947; Friberg, 1950; Apostolov, 1979) and anosmia (Friberg, 1950). Anosmia was found by Friberg (1950) in about one third of a group of workers with a mean exposure time to cadmium oxide dust of 20 years. Baader (1951) in Germany and Apostolov (1979) in Bulgaria also noted that anosmia was common among workers exposed to cadmium oxide dust for long periods of time. Suzuki et al. (1965) and Tsuchiya (1967) in Japan, found no increase in the prevalence of anosmia in workers exposed to cadmium stearate and cadmium oxide fumes.

Nervous system symptoms were reported by Vorobjeva (1957), who investigated 160 workers at an accumulator factory in the

USSR. Subjective symptoms included headache, vertigo, and sleep disturbance. Physical examination revealed increases in knee-joint reflexes, tremor, dermographia, and sweating.

Cadmium sulfide is sometimes used as a yellow tattoo pigment, which is deposited intradermally. Local phototoxic reactions may take place when the skin is exposed to ultraviolet light and are probably connected with the marked photoconducting properties of cadmium sulfide. Of 24 patients with yellow tattoos who were examined by Bjornberg (1963), 18 experienced skin swelling when exposed to sunlight.

## 8.3 Clinical and epidemiological studies with data on both exposure and effects

There are several clinical and epidemiological studies with data on occupational or general environment exposure levels, but the data concerning effects are restricted to the lungs and kidneys.

### 8.3.1 *Studies on respiratory disorders*

Friberg (1950) studied 43 male workers exposed to cadmium oxide dust, with an average period of employment of 20 years (range 9-34 years), and 15 workers who had been employed for only 1-4 years. They were compared with a group of 200 sawmill workers. Shortness of breath was the common symptom among the workers with long exposure, and an impairment of lung function (increased residual capacity in relation to total lung capacity and a decreased working capacity) was demonstrated. The lung function of the group with short exposure (less than 5 years) was found to be normal. The cadmium concentration in air varied from 3 to 15 $mg/m^3$, measurements having been made at five places on only one occasion. In another battery factory (where air cadmium concentrations were 0.05-5 $mg/m^3$, Adams et al. (1969) found a slight average decrease in forced expiratory volume in a group of 27 male workers.

Twelve out of 96 cadmium workers exposed for up to 27 years to cadmium oxide fume in two cadmium-copper alloy factories were found to suffer from emphysema, as evaluated from a comprehensive lung function test (Bonnell, 1955; Buxton, 1956; Kazantzis, 1956). These workers were compared with a similar size control group. The average air cadmium concentrations in the two factories were 40-50 $\mu g/m^3$, and 90% of the particles were less than 0.5 $\mu m$ in diameter (King, 1955).

Lauwerys et al. (1979a) studied pulmonary ventilatory function in three groups of workers exposed to cadmium oxide dust and in matched control groups (the matching included smoking habits). A slight but significant reduction in forced vital capacity, in forced expiratory volume in one second, and in peak expiratory flow rate was found in 22 men. These were all smokers and had been exposed for more than 20 years to a time-weighted average air concentration of 66 $\mu g/m^3$ (21 $\mu g/m^3$ respirable cadmium). In another group of workers (smokers and non-smokers) exposed for 1-20 years to an average concentration of 134 $\mu g/m^3$ (the respirable cadmium level at the most polluted work site was 88 $\mu g/m^3$), the pulmonary indices were on average lower than in the control group, but the differences were not statistically significant. A more thorough examination of a subgroup of the workers with long exposures carried out by the same research group (Stanescu et al., 1977) found more respiratory symptoms in the cadmium-exposed group than in a control group and also impaired lung function (not statistically significant). However, Lauwerys et al. (1979a) reported more extensive data from the same plants and found that the workers with less than 20 years of exposure (average 7.5 years) showed significant effects in the lung function tests.

Reduced forced vital capacity was also found at a cadmium production plant in the USA (where the air concentrations were "commonly greater than 200 $\mu g$ $Cd/m^3$") among 17 workers exposed for more than 6 years (Smith et al., 1976). De Silva & Donnan (1981) provided evidence that insoluble cadmium compounds may induce emphysematous changes after more than 7 years exposure to a time-weighted average cadmium concentration of 700 $\mu g/m^3$.

Edling et al. (1986) found no lung function differences between an exposed group of workers using cadmium-containing solders and a control group. The level of exposure, which lasted for several years, was estimated to be 0.05-0.5 mg cadmium/$m^3$, but the workers had not been exposed to cadmium for several years. Of the 57 workers examined, 42% had cadmium-induced renal tubular dysfunction.

Davison et al. (1988) examined 101 male workers, who had worked for 1 year or more manufacturing copper-cadmium alloys, and found, compared with a reference group, impaired lung function. They also compared certain parameters (transfer coefficient: KCO) with the estimated cumulative exposure index for cadmium workers with 95% confidence limits for the regression line. Among 35 workers exposed for more than five years and with a cumulative cadmium exposure index up to 14 mg/m$^3$.years, there was no evidence of a threshold. The authors concluded that inhaled cadmium fumes caused changes in lung function and in chest radiographs consistent with emphysema. This could also explain the increased mortality reported. The impaired lung function was also related to liver cadmium levels as measured with neutron activation *in vivo*.

Some studies on respiratory disorders have yielded negative results. However, some of these studies did not use lung function tests (Hardy & Skinner, 1947; Princi, 1947; Tsuchiya, 1967; L'Epee et al., 1968) and another did not use a control group (Teculescu & Stanescu, 1970). Suzuki et al. (1965) examined a group of workers exposed for a short period (average 3.3 years) to 30-690 $\mu$g cadmium/m$^3$ (as cadmium stearate) and found no changes in lung function when compared to a control group.

In summary, it is clear that exposure to cadmium dust and fume over prolonged periods can give rise to impaired lung function and emphysema. Such effects have been seen predominantly at high air cadmium concentrations (above 100 $\mu$g/m$^3$), but one study showed effects after more than 20 years of exposure to respirable cadmium oxide dust concentrations of 21 $\mu$g/m$^3$.

Cadmium workers sometimes suffered from symptoms such as coughing and throat irritation, but did not show abnormal chest X-ray findings when exposed to cadmium oxide fume at a concentration of 100 $\mu$g/m$^3$ for 4-8 years (Hardy & Skinner, 1947) or 40-1440 $\mu$g/m$^3$ for 8 years (Princi, 1947), or to cadmium oxide dust at a concentration of 64-241 $\mu$g/m$^3$ for up to 15 years (Tsuchiya, 1967).

### 8.3.2 *Studies on renal disorders in industry*

Friberg (1950) reported that prolonged exposure gave rise to renal damage among a large group of workers exposed to cadmium oxide dust at concentrations of 3-15 mg/m$^3$ in an accumulator factory. In one group of 43 workers with a mean exposure period of 20 years (range 9-34 years), a high prevalence of proteinuria was demonstrated by the nitric acid and trichloroacetic acid test. In several of the workers, the renal damage was also manifested by a decreased inulin clearance and decreased concentrating capacity.

Another group of 15 workers with a mean exposure period of 2 years (range 1-4 years) showed no positive reactions.

Since 1950, there have been many studies on proteinuria among workers in various industries. This type of proteinuria is characterized particularly by a great relative increase in the excretion of low molecular weight (LMW) proteins (section 8.2). In most of the early studies, qualitative tests for detecting proteinuria were used, but, more recently, specific methods for the quantitative determination of LMW proteins have been developed.

Table 16 contains data on the prevalence of proteinuria from several epidemiological studies on cadmium workers. It must be recognized that, in most studies, the dose measurements are based on short sampling periods (hours or a few days), whereas exposure may have been for decades. Information on sampling method (static or personal) and the use of respirators is usually inadequate, which makes accurate dose estimates difficult (section 2.2.1).

It is evident that the prevalence of proteinuria in cadmium workers increases with exposure intensity duration. The study by Kjellström et al. (1977a) presents frequency distributions of urinary $\beta_2$-microglobulin excretion for 240 exposed workers and a control group. There is a general shift to higher excretion levels among the exposed workers, and a large proportion of them have excretion levels far outside the control distribution. Any cut-off point (operational definition) for "abnormal" proteinuria is arbitrary. If a cut-off point of 290 $\mu$g/litre (corresponding to the 97.5 percentile of the control group) is chosen, 26% of the whole group of exposed workers would be classified as having LMW proteinuria. If higher cut-off points are used, the prevalence of proteinuria will obviously be lower.

Table 16 provides evidence of a dose-response relationship. The lowest "dose" that gave rise to a statistically significant increase in urinary $\beta_2$-microglobulin, as defined above, was a 6- to 12-year exposure to 50 $\mu$g cadmium/m$^3$ (based on personal sampling) (Kjellström et al., 1977a).

Järup et al. (1988) recently made a reassessment of dose-response in the same battery plant (Table 16). A pattern very similar to Kjellström et al. (1977a) was observed with a prevalence of $\beta_2$-microglobulinuria of 4% at a cumulative dose of 0.5 mg/m$^3$ (corresponding to 10 years of exposure to 50 $\mu$g cadmium/m$^3$).

Lauwerys et al. (1979a,b) studied the prevalence of increased $\beta_2$-microglobulin clearance (cut-off point: 97.5 percentile of controls) and found a 21% prevalence after more than 20 years of exposure to 66 µg cadmium/m$^3$ total dust (static samples) or 21 µg cadmium/m$^3$ respirable dust (Table 16).

Holden (1980b) measured urine levels of $\beta_2$-microglobulin and found dose-response relationships using cut-off points of 200 µg/litre, 1000 µg/litre, or 10 000 µg/litre. The cut-off point of 200 µg/litre gave a 16% LMW proteinuria prevalence rate after 6-10 years of exposure (Table 16).

Table 16 also shows that an increased prevalence of total proteinuria, as measured by sulfosalicyclic acid, trichloroacetic acid, or quantitative determination of total proteinuria, occurs after 5-10 years exposure to approximately 100 µg cadmium/m$^3$. An increased excretion of LMW protein (e.g., $\beta_2$-microglobulin) occurs at much lower doses.

The *in vivo* measurement of cadmium in the liver and kidneys of people with various levels of cadmium exposure provides a means for relating organ dose to effects and response rates (section 6.4.2). Some questions still remain regarding the accuracy of the analytical method (section 2.2.3.4) and the mathematical-statistical methodology (Kjellström et al., 1984). Nevertheless, Roels et al. (1983a) and Ellis et al. (1984) suggested that renal tubular damage would be experienced by about 10% of people with a kidney cortex level of 200 mg cadmium/kg, and by about 50% of people with a kidney cortex level of 300 mg cadmium/kg.

Ellis et al. (1985) correlated time-weighted exposure indices (TWE), based on employment records, area monitoring techniques and personal sampling, with body burden of cadmium measured by *in vivo* neutron activation analysis of the liver and left kidney in 82 men exposed to cadmium dust in a smelter. The workers were grouped as follows: production workers (40 active, 21 retired); office and laboratory workers (8 active, 4 retired); and non-production workers (3 active, 6 retired). From these measurements the authors were able to estimate the probability of developing kidney dysfunction based on the workers' cumulative exposure index. When the exposure limit was 400-500 µg/m$^3$.years, the prevalence for renal dysfunction was about 32%; it was 22-40% with a wider exposure index (300-600 µg/m$^3$). Kjellström et al. (1977a) reported a prevalence of 19% at a battery factory at a level of 50 µg Cd/m$^3$, which resulted in a similar

Table 16. Prevalence of proteinuria in cadmium workers

| Cadmium compounds | Estimated air concentrations ($\mu g/m^3$)[a] | Exposure period (years)[b] | No. of examinees | Prevalence of proteinuria (%) | Proteinuria and characteristics of detection method[e] | Reference |
|---|---|---|---|---|---|---|
| Cadmium oxide fume | 40-50 | control | 60 | 2 | SA and TCA | Bonnell (1955); King (1955); Bonnell et al. (1959) |
| | | 1-9 | 37 | 24 | | |
| | | > 9 | 63 | 46 | | |
| | 64-241[c] | control | 11 | 0 | TA | Tsuchiya (1967) |
| | | < 1 | 4 | 0 | < 100 mg/litre | |
| | | 1-4 | 4 | 50 | | |
| | 123[c] (time-weighted average | > 5 | 4 | 100 | | |
| Cadmium oxide dust | 3000-15 000 | 1-4 | 15 | 0 | nitric acid ("Hellers test"); positive in more than half the test | Friberg (1950) |
| | | 9-15 | 12 | 33 | | |
| | | 16-22 | 17 | 41 | | |
| | | 23-34 | 14 | 64 | | |
| | 31 (1.4)[c] | control | 31 | 0 | Abnormal electrophoretic pattern as defined by the authors; $\beta_2$-microglobulin clearance | Lauwerys et al. (1974a) |
| | | 1-12 (4) | 31 | 0 | | |
| | | control | 27 | 4 | | |
| | 134 (88)[c] | 0.6-19.7 (9) | 27 | 15 | | |
| | | control | 22 | 0 | | |
| | 66 (21) | 21-40 | 22 | 68 | | |
| | 66 (21) | > 20 | 42 | 21 | | |

Table 16 (contd).

| | | | | | | |
|---|---|---|---|---|---|---|
| Cadmium oxide dust | | | | | | |
| | | control | 87 | 3.4 | $\beta_2$-microglobulin | Kjellström et al. (1977a) |
| | 50[c] | 0-3 | 50 | 6.0 | (RIA) | |
| | 50[c] | 3-6 | 30 | 6.6 | > 290 $\mu$g/litre | |
| | 50[c] | 6-12 | 21 | 19.0 | (sg = 1.023) | |
| Cadmium stearate dust | 30-690 | control | 24 | 17 | TCA | Suzuki et al. (1965) |
| | | (3) | 19 | 58 | | |
| Cadmium sulfide dust | 114[d] | < 1-5 | 12 | 17 | EP | Harada (1987) |
| | | 5-21 | 7 | 100 | | |
| | | < 1-5 | 12 | 8 | TCA | |
| | | 5-21 | 12 | 43 | | |
| | | control | 203 | 1 | $\beta_2$-microglobulin | Stewart & Hughes (1981) |
| | 80 | 0-5 | 105 | 0 | (RIA) | |
| | 100 | 6-11 | 41 | 0 | > 765 $\mu$g/litre | |
| | | | | | (sg = 1.016) | |
| | 100-600 | 11-19 | 13 | 7.7 | | |
| | 100-600 | 20+ | 14 | 57.0 | | |
| Mixed exposures; | not reported, | controls | 642 | 0 (0.8) | $\beta_2$-microglobulin | Holden (1980b)[f] |
| 12 factories; | but mean blood | < 18 months | 121 | 0 (10) | (RIA) | |
| mainly zinc | cadmium level | 19 months-5 years | 168 | 1.8 (8.3) | > 1000 $\mu$g/litre | |
| smelters | after 1 year | 6-10 | 170 | 1.8 (16) | (or > 200 $\mu$g/litre) | |
| | exposure was | 11-15 | 82 | 7.3 (22) | | |
| | about 15 $\mu$g/litre | 16-20 | 33 | 24 (45) | | |
| | | 20+ | 68 | 25 (56) | | |

Table 16 (contd).

| Cadmium compounds | Estimate of cumulative dose ($mg/m^3$.year) | Number of examinees | Prevalence (%) of proteinuria | Proteinuria and characteristics of detection method | Reference |
|---|---|---|---|---|---|
| Cadmium fume (oxide) | < 1 | 16 | 18 | $\beta_2$-microglobulin > 0.034 mg/mmol creatinine | Elinder et al. (1985b) |
| | 1-2 | 22 | 32 | | |
| | 2-3 | 9 | 44 | | |
| | 3-5 | 8 | 62 | | |
| | > 5 | 5 | 100 | | |
| | < 1 | | 6 | | Mason et al. (1988) |
| | 1-15 | | 67 | | |
| | 1.5-3 | 75 | 58 | | |
| | 3-5 | | 86 | | |
| | > 5 | | 100 | | |
| Cadmium oxide | < 0.4 | 264 | 1.1 | | Jarup et al. (1988) |
| | 0.4-1.7 | 76 | 9.2 | | |
| | 1.7-4.6 | 43 | 23.3 | | |
| | 4.6-9.6 | 31 | 32.3 | | |
| | 9.6-15 | 16 | 32.2 | | |
| | > 15 | 10 | 50 | | |

[a] Numbers in parentheses refer to average concentrations of the respirable particulates (< 5 μm) fraction

[b] Numbers in parentheses are average values

[c] Measured in breathing zone by personal sampler

[d] Calculated average exposure for the worker with the most pronounced effect

[e] SA = sulfosalicyclic acid method (qualitative); TCA = trichloroacetic acid method (qualitative); TA = tungstate method (quantitative); EP = electrophoresis; RIA = radioimmunoassay; sg = specific gravity

[f] Values for prevalence of proteinuria refer to a $\beta_2$-microglobulin level of > 1000 μg/litre. Values in parentheses refer to a level of > 200 μg/litre.

exposure index. Lauwerys et al. (1974a) observed proteinuria in 68% of workers with long-term exposure (20 years at 66 $\mu g/m^3$), whereas the logistic model developed by Ellis et al. (1985) would have predicted 65% under these exposure conditions.

For exposures of 250 $\mu g/m^3$.years there is a 19% probability of experiencing renal dysfunction. In the case of workers with normal renal function, an exposure index of 400 $\mu g/m^3$ predicts a mean cadmium concentration of 28 mg/kg in liver and 288 mg/kg of renal cortex. The model would predict that a 10-year exposure to 25 $\mu g/m^3$ would result in a mean renal cortex concentration of 252 mg/kg, which is similar to the critical concentration defined by Friberg.

Thun et al. (1989) assessed the quantitative relationship between exposure to airborne cadmium and various markers of renal tubular and glomerular function in 45 male workers at a plant that recovered cadmium from industrial waste. The dose was estimated from historical air sampling data. In this study, the "critical dose" of cadmium necessary to induce nephropathy was based on the 5th or 95th percentile of test results in the unexposed population. Using this definition, renal tubular dysfunction sharply increased as cumulative exposure to cadmium rose above 300 $mg/m^3$.days (corresponding to about 0.8 $mg/m^3$.years). Very similar dose-response curves with an increased prevalence of $\beta_2$-microglobulinuria at cumulative exposure levels exceeding about 0.5-1 mg cadmium/$m^3$.year have been reported from examinations of workers exposed to cadmium fumes (Elinder et al., 1985b; Mason et al., 1988). These findings are consistent with the recommendations by a working group of the World Health Organization (WHO, 1980) to limit workplace exposures to 10 $\mu g/m^3$ in order to prevent tubular proteinuria after life-time occupational exposure to cadmium.

Cumulative cadmium exposure indices have been calculated for 75 cadmium alloy workers employed for periods of up to 39 years, together with the *in vivo* liver and kidney cadmium burden (Mason et al., 1988). Several indicators of both tubular and glomerular dysfunction correlated significantly with both cumulative exposure index and liver cadmium burden. Using these estimates of dose, a two-phase linear regression model was applied to identify an inflection point of the order of 1100 $\mu g/m^3$.years above which changes in renal function occurred. A number of biochemical variables fitted this model, including total protein, albumin, and $\beta_2$-microglobulin. Simple dose-response analysis showed a greatly increased incidence of tubular proteinuria when the cumulative cadmium exposure index was greater than this value. The cumulative exposure index was equated to about 20 to 22 years of exposure to a cadmium level of 50 μg/m$^3$.

Evidence of tubular damage was investigated in a group of 91 cadmium workers subjected to yearly estimation of cadmium concentration in blood and urine over a period of eight years. In workers with blood and urine cadmium levels constantly below the Biological Limit Value of 10 μg/litre, the prevalence of tubular damage, as indicated by an increased excretion of $\beta_2$-microglobulin above 260 μg/litre, was below 3%. RBP excretion confirmed this pattern.

### 8.3.3 Studies on renal disorders in the general environment

#### 8.3.3.1 Health surveys in Japan

Following the recognition of the association between Itai-itai disease and exposure to cadmium (Japanese Ministry of Health and Welfare, 1968), additional general surveys of cadmium pollution were performed in Japan, and further areas were found to be involved. Health effects were studied first among the population in the area where Itai-itai cases had occurred and later in other areas found to be contaminated. The original studies were designed to find cases of Itai-itai disease, but it was possible also to estimate the prevalence of proteinuria and glucosuria in the examined population. A detailed description of the methods used in these cadmium pollution surveillance programmes has been reported by Shigematsu et al. (1979).

At the time of the first surveillance programmes (1969-71), methods were developed for estimating the degree of contamination with cadmium and the total daily intake. At the early stage of the investigations, the most common index of cadmium intake measured in all areas was the cadmium concentration in rice. From data of the Japan Public Health Association (1970), it was estimated that, in areas with different exposure levels, an average of almost 50% (range 14-71) of the daily cadmium intake came from rice. In one area, the proportion was estimated to be 85% (Kawano & Kato, 1978).

The average cadmium concentration in rice from non-polluted areas has been reported to be 0.066 mg/kg in polished rice

(Moritsugi & Kobayashi, 1964) and 0.09 mg/kg in unpolished rice (Japanese Ministry of Agriculture and Forestry, 1973) (section 5.2.1). The national average consumption of rice was 364 g per person in 1961, 308 g in 1971, and 222 g in 1981 (Japanese Ministry of Health and Welfare, 1983).

Proteinuria was generally estimated with qualitative methods such as the sulfosalicylic acid method or the trichloracetic acid method according to standardized techniques (Japanese Ministry of Health and Welfare, 1971). More recently, emphasis has been placed on the identification of the urinary protein pattern, in particular to detect early evidence of tubular dysfunction. The methods currently used are electrophoresis and the quantitative determination of lysozyme, RBP, and $\beta_2$-microglobulin. Several studies have been published, and there are extensive reviews in English (Tsuchiya, 1969; Yamagata & Shigematsu, 1970; Friberg et al., 1974; Tsuchiya, 1978; Shigematsu et al., 1979, 1980).

From 1976 to 1984, epidemiological health surveys of residents in areas with environmental cadmium pollution were performed by the Japan Environment Agency using methods including immunological tests for the detection of low molecular weight proteinuria in eight prefectures (Akita, Fukushisuma, Gunma, Toyama, Ishikawa, Hyogo, Nagasaki, Oita). More than 13 000 inhabitants of polluted areas and more than 7000 inhabitants of non-polluted areas, aged 50 years or more in both areas, were subjected to these surveys.

The following screening method was adopted for health examinations. The urine of those people with proteinuria (demonstrated by a semiquantitative method) and/or glucosuria (by a paper test) was analysed for $\beta_2$-microglobulin ($\geq$ 10 mg/litre), RBP ($\geq$ 4 mg/litre), lysozyme ($\geq$ 2 mg/litre), total amino acid nitrogen ($\geq$ 20 mmol/litre), and cadmium ($\geq$ 30 $\mu$g/litre). Those who exceeded the above levels in more than one item were tested for renal function by urine and blood analysis. Finally those for whom the TRP was less than 80% were subjected to detailed health examination, including skeletal radiography, in order to make a clinical diagnosis (Fig. 7).

With the exception of Oita prefecture, the number of individuals who had or were suspected of having proximal renal tubular dysfunction (as defined by the Japanese Cadmium Research Committee) or related findings tended to be greater in the polluted areas than in the non-polluted areas, and this was

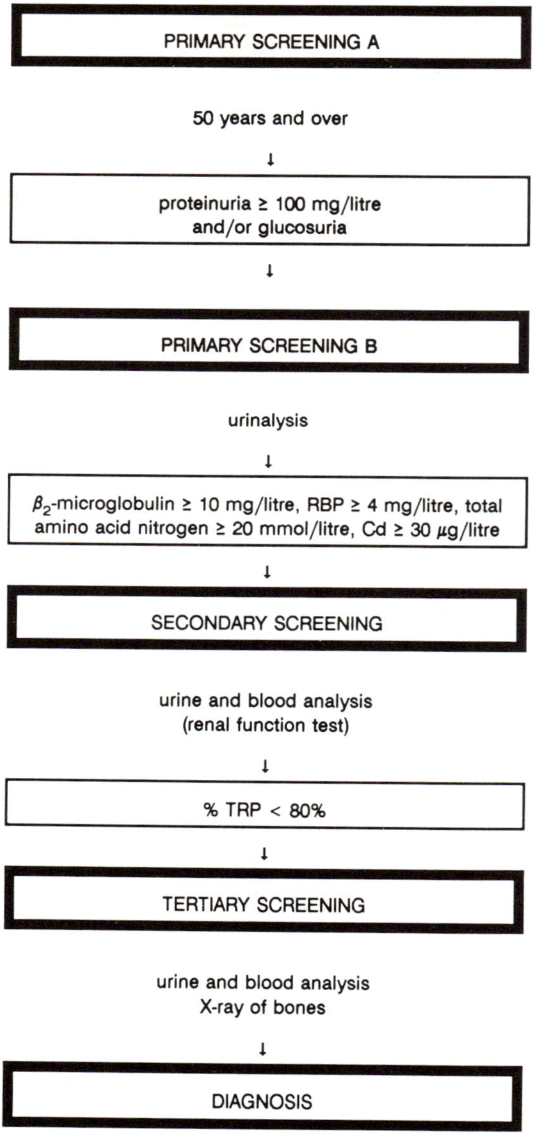


Fig. 7. The revised method of health examination in Japanese cadmium-polluted areas From: Shigematsu et al. (1980).

often significantly related to the degree of pollution (see Table 16). This suggests that environmental cadmium pollution is associated with the occurrence of proximal renal tubular dysfunction.

Five areas in which significantly increased $\beta_2$-microglobulinuria was found are reviewed in detail below. In addition there is a description of some other Japanese polluted areas and three European polluted areas.

### 8.3.3.2 *Toyama prefecture (Fuchu area)*

This is the area where the Itai-itai disease was first described (Kono et al., 1956). Exposure levels in polluted villages, as measured by cadmium concentrations in rice during the 1960s, varied greatly but in some villages the level was as high as 2 mg/kg (Ishizaki et al., 1969). A zinc and lead mine was the major source of pollution, and cadmium concentrations in soil were elevated (Japan Public Health Association, 1968). Many studies have been performed with sulfosalicyclic acid and trichloroacetic acid for the identification of proteinuria. Both proteinuria and glucosuria were common findings in the polluted area (Ishizaki et al., 1969; Fukushima et al., 1974). There was a strong relationship between the degree of proteinuria and age, and a greatly increased prevalence in the older age groups compared with controls. The proteinuria was similar to that seen in cadmium workers as evaluated by electrophoresis (Piscator & Tsuchiya, 1971) or gel filtration (Fukuyama, 1972). Quantitative estimation of the LMW urinary proteins $\beta_2$-microglobulin (Shiroishi et al., 1975, 1977) and retinol-binding protein (Kanai et al., 1971) confirmed that the proteinuria was tubular.

Fukushima et al. (1973) reported on the cadmium concentration in rice and the prevalence of renal effects in various hamlets in the Fuchu and control areas. In control hamlets situated outside the Jinzu river basin, the cadmium concentration in rice varied between 0.05 and 0.2 mg/kg wet weight, and the prevalence of concurrent proteinuria and glycosuria varied between 0 and 9%. In the polluted villages, where cadmium levels in polished rice were 0.5-1.0 mg/kg, the prevalence was 15-20%, and, in all the 20 hamlets, the correlation coefficient between cadmium in rice and prevalence of renal effects was 0.68 ($P < 0.05$) (Fukushima et al., 1973). The prevalence had a tendency to be somewhat higher in the hamlets where Itai-itai disease was endemic, as compared with the hamlets where it did not occur, even though the latter hamlets had similar cadmium concentrations in rice.

Using a disc electrophoresis technique (Shiroishi et al., 1972) in the age groups over 40 years, a tubular urinary protein pattern was found in about 25% of exposed persons but not found at all in the control group. Quantitative determination of $\beta_2$-microglobulin in the urine of patients with Itai-itai disease and so-called observation patients (people in polluted areas with likely cadmium-induced renal damage) showed a marked difference between the exposed and control groups (Shiroishi et al., 1977).

The level of urinary $\beta_2$-microglobulin in patients with Itai-itai disease was on average 43 mg/litre, i.e. 100 times higher than in the controls, and the level in observation patients was on average 65 times higher than it was in the controls.

A comparison by Kjellström et al. (1977b) of 138 cadmium-exposed women in the age-group 51-60 and 40 controls revealed large differences. The exposed women were selected on the basis of their consumption of polluted rice (average cadmium concentration above 0.7 mg/kg); no health data influenced the selection. On average, the urinary $\beta_2$-microglobulin excretion was 10 times greater among the exposed women than among the controls, and the individual urinary levels increased as a function of the cadmium dose. Additional data from the same area for women in the age-group 40-45 (Kjellström, 1977) also showed an increase prevalence of high $\beta_2$-microglobulin levels in urine.

Nogawa & Ishizaki (1979) reported a significant increase in the prevalence of both proteinuria and concurrent proteinuria and glucosuria at an average rice cadmium level of 0.41 mg/kg. In a further study (Nogawa et al., 1979), the prevalence of proteinuria was analysed as a function of urinary cadmium levels. The correlation between the two variables was good, but urinary cadmium may not be a suitable measure of dose as it also increases as a consequence of renal damage (section 8.2.1).

A mathematical dose-response analysis was carried out by Hutton (1983) based on the data of Shiroishi et al. (1977) on urinary $\beta_2$-microglobulin excretion. For each individual the cadmium dose index ($\mu$g/day.years) was based on the estimated daily cadmium intake via rice and other foodstuffs and the number of years the person had lived in the polluted area. The age-groups 51-60 and 40-45 were both divided into three sub-groups with different dose index levels. In the analysis of Hutton (1983), three groups from Kosaka area were included (section 8.3.3.4) for whom the same type of data was available (Kojima et al., 1977). Fig. 8 shows that the prevalence of increased LMW

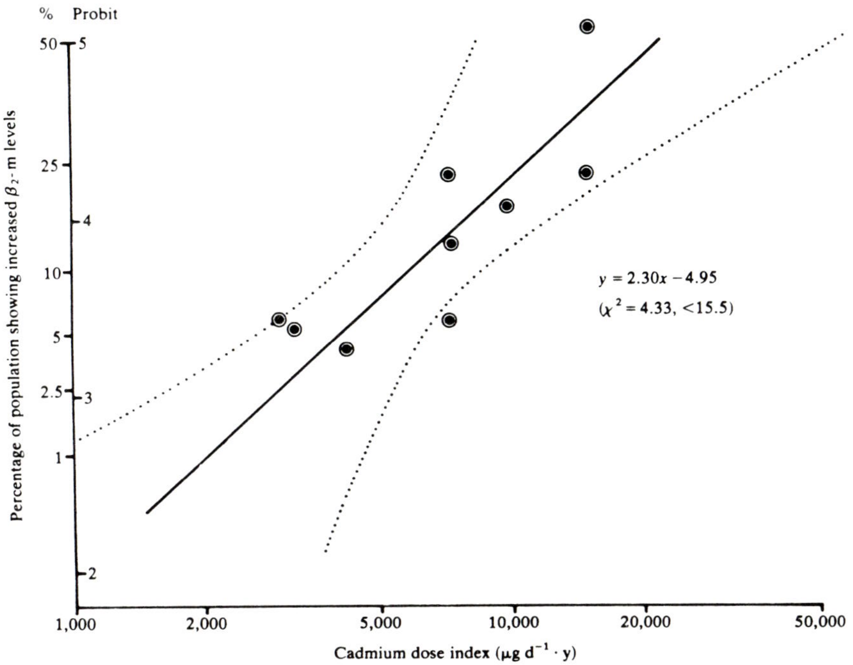


Fig. 8. Dose-response relationship for cadmium-induced increases in urinary $\beta_2$-microglobulin level (μg/g creatinine), with upper and lower 95% fiducial limits. From: Hutton (1983), based on data from Kjellström (1977)

proteinuria (response rate) increased with dose. These prevalences were adjusted for a control group prevalence of 2.5% (Kjellström, 1977), giving an expected "background" prevalence without cadmium exposure of 0%. The 95% fiducial limits were quite wide. For instance, at a cadmium intake of 55 μg/day (95% fiducial limits 25-123), there would be a 1% increase of proteinuria in the population. At a dose index of 5000 μg/day.years (or 50 years at 100 μg cadmium/day intake), the expected response rate was within the range 2-12%.

#### 8.3.3.3 *Hyogo prefecture (Ikuno area)*

In the Ikuno area of Hyogo prefecture, an inactive zinc and copper mine is the probable source of pollution of the Ichi river basin. The average cadmium concentration in rice in the most polluted part was found to be 0.69 mg/kg in one study and 1.10 mg/kg in another (Hyogo Prefectural Government, 1972).

In 1972, urine from 1560 people (of both sexes, over 30 years of age) from the polluted area and groups of 1574, 2002, and 638 people (over 30) from three control areas were examined (Tsuchiya, 1978). The prevalence of proteinuria, as measured by the sulfosalicyclic acid methods was 58% and 33%, respectively, a statistically significant difference. The reason for the high prevalence in the control area is not known.

In a study by Watanabe & Murayama (1975), a search was made for LMW proteins among 39 people in a polluted area and 56 in a control area (all the people were above 70 years of age). Urinary $\beta_2$-microglobulin excretion exceeding 10 000 μg/litre was found in 41% of the examined people in the polluted area and 4% of those in the control area.

Kitamura & Koizumi (1975) used disc electrophoresis to study tubular-type proteinuria among 224 people (above 50 years of age) from a polluted area and compared the results with those from a study of old bedridden people. Fig. 9 demonstrates the considerably higher prevalence of tubular proteinuria among people from the polluted area. There was also a definite increase in the occurrence of tubular proteinuria with age in the exposed and bedridden control groups.

#### 8.3.3.4 *Ishikawa prefecture (Kakehashi area)*

In the Kakahaski river basin of Ishikawa prefecture, several mines had polluted the river with cadmium and copper (Tsuchiya,

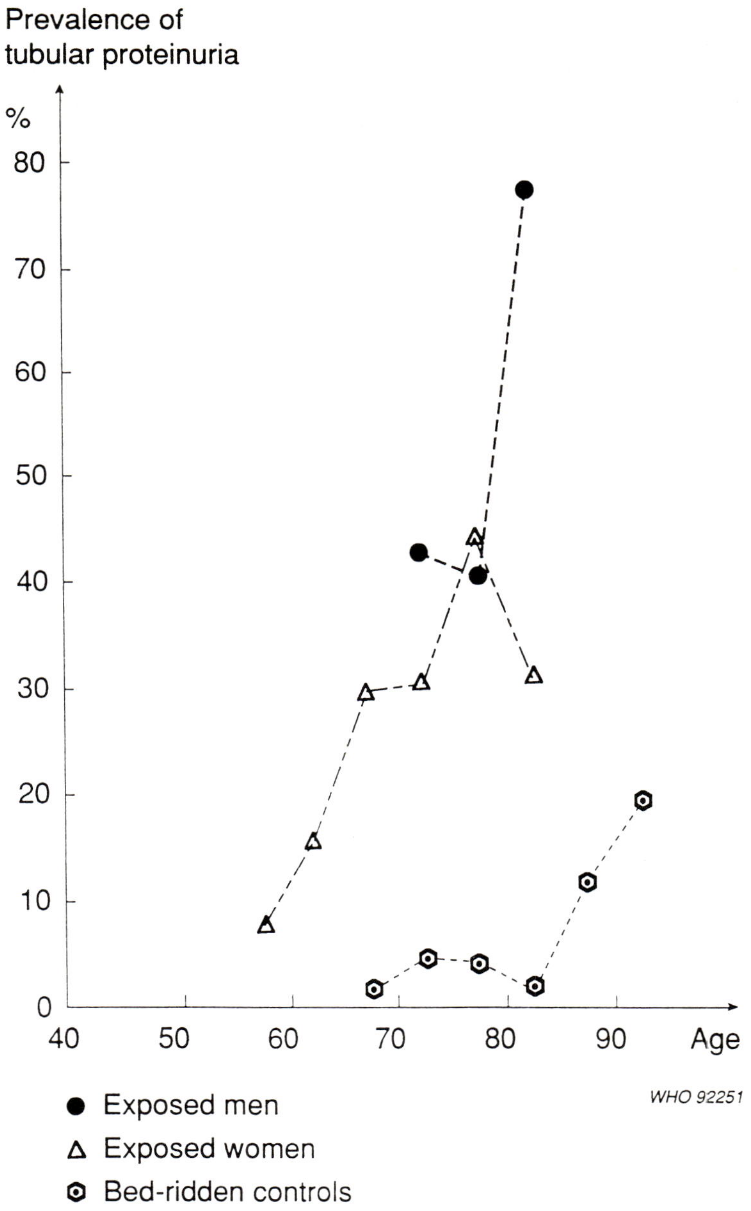


Fig. 9. Prevalence of tubular proteinuria among cadmium-exposed people and old people confined to bed in Hyogo prefecture From: Kitamura & Koisumi (1975).

1978). Rice samples were studied in a number of villages along the river, and village-average levels of up to about 0.7 g/kg were found. Values were higher in paddy fields on the shores of a narrow river valley close to the mine.

In 1974-1976, health examinations of 2805 inhabitants over the age of 50 were carried out using test tape for proteinuria and glucosuria examination as well as single radial immunodiffusion analysis for RBP in urine (Tsuchiya, 1978). Based on the findings, some people were selected for secondary and tertiary examinations, and 39 were considered to require consultation for renal tubular dysfunction. However, no cases with severe bone disease were found at that time.

This area is the only one where quantitative measurement of LMW protein in urine was carried out in the first screening (Nogawa et al., 1978). The prevalence data in Table 17 are therefore of particular interest. None of the other LMW proteinuria studies mentioned in Table 16 were carried out on such a large group using a broad epidemiological approach. A scatter in the prevalence values among the villages is seen in Fig. 10, but there is no doubt that the villages with the highest rice cadmium values had an increased prevalence of high urinary RBP. This is also evident when the data from different villages with the same rice cadmium levels are combined (Table 17). In all of the exposed groups with different rice cadmium levels (Table 17), the prevalence of tubular proteinuria increases with age, but that is not seen in the control group. It is not known whether the age effect reflects increased cadmium dose rather than age itself.

Further analysis of these data using a mathematical dose-response approach (Hutton, 1983) clearly showed the effect of calculated cadmium intake on the prevalence of proteinuria (Fig. 11). The fiducial limits are narrower than in Fig. 8 because of the larger number of people studied.

In a study by Nogawa et al. (1978), laboratory determinations related to proximal renal tubular function, etc., were compared by age group. The findings in the most polluted areas hardly differed from those in the non-polluted areas in the age-group 50-59 (Table 17). At age 60 and over, however, the frequency of findings tended to increase with age, except in the case of total aminoaciduria. The difference in the age-adjusted rates for these determinations between the polluted and non-polluted areas tended to be enhanced by aging. Table 18 shows how the

Table 17. Age-adjusted prevalence rate (%) of renal tubular dysfunction and related conditions[a]

| Prefecture studied[b] | Year of investigation | Polluted (P) or non-polluted (NP) area[c] | No. of examinees | | $\beta_2$-Microglobulinuria (> 10 mg/litre) | | Decreased TRP (< 80%) | | Tubular dysfunction | |
|---|---|---|---|---|---|---|---|---|---|---|
| | | | male | female | male | female | male | female | male | female |
| Toyama | 1979-1984 | P | 3432 | 4099 | 6.5 | 10.8 | 4.6 | 5.5 | 1.4 | 3.3 |
| (Fuchu area) | | NP | 944 | 1205 | 0.4[d] | 0.5[d] | 0.6[d] | 0.2[d] | 0.0[d] | 0.0[d] |
| Hyogo | 1977 | P | 230 | 280 | 12.8 | 16.8 | 4.4 | 6.8 | 2.0 | 3.5 |
| (Ikuno area) | | NP | 212 | 251 | 2.7[d] | 1.9[d] | 0.9[e] | 0.4[d] | 0.0 | 0.0[e] |
| Ishikawa | 1976 | P | 260 | 306 | 7.6 | 10.9 | 6.9 | 5.7 | 2.4 | 3.2 |
| (Kakehashi area) | | NP | 200 | 275 | 1.1[d] | 1.5[d] | 0.5[d] | 0.0[d] | 0.0 | 0.0[d] |
| Akita | 1976 | P | 179 | 247 | 6.4 | 5.0 | 2.3 | 0.7 | 0.0 | 0.0 |
| (Kosaka area) | | NP | 168 | 234 | 0.0[d] | 0.0[d] | 0.0 | 0.0 | 0.0 | 0.0 |
| Nagasaki | 1976 | P | 143 | 191 | 3.4 | 10.6 | 4.6 | 9.1 | 1.7 | 6.2 |
| (Tsushima area) | | NP | 210 | 291 | 1.9 | 0.3[d] | 1.5 | 0.0[d] | 0.0 | 0.0[d] |
| Fukushima | 1977 | P | 307 | 425 | 0.4 | 0.5 | 0.6 | 0.5 | 0.0 | 0.3 |
| | | NP | 246 | 396 | 0.0 | 0.0 | 0.4 | 0.0 | 0.0 | 0.0 |

Table 17 (contd).

| Prefecture studied[b] | Year of investi-gation | Polluted (P) or non-polluted (NP) area[c] | No. of examinees | | $\beta_2$-Microglobulinuria (> 10 mg/litre) | | Decreased TRP (< 80%) | | Tubular dysfunction | |
|---|---|---|---|---|---|---|---|---|---|---|
| | | | male | female | male | female | male | female | male | female |
| Gunma | 1976-1978 | P | 937 | 1160 | 1.6 | 0.6 | 1.6 | 0.5 | 0.0 | 0.0 |
| | | NP | 620 | 786 | 0.8 | 0.1 | 0.6 | 0.3 | 0.0 | 0.0 |
| Oita | 1978 | P | 169 | 194 | 1.7 | 1.5 | 1.1 | 0.5 | 0.0 | 0.0 |
| | | NP | 182 | 215 | 1.7 | 1.0 | 1.2 | 0.0 | 0.0 | 0.0 |

[a] From: Japan Cadmium Research Committee (1989). The response rates for these studies were greater than 90% of the target population. The age composition (50-59, 60-69, 70-79, and 80+) in each prefecture was adjusted to the Japanese population in 1980. The criteria for renal tubular dysfunction were: one out of three signs (low molecular weight proteinuria, glucosuria, and generalized aminoaciduria), %TRP ≤ 80% and acidosis (blood hydrogen carbonate below 23 mEq/litre).

[b] The total number of people living in polluted areas in each prefecture is shown in Table 7.

[c] Total number of people examined was 5657 males and 6902 females in polluted areas and 2782 males and 3653 females in non-polluted areas.

[d] Significant difference ($P < 0.01$)

[e] Significant difference ($P < 0.05$)

Table 18. Prevalence (%) of tubular proteinuria[a] in relation to age (age-groups 50-59, 60-69, and > 69) and village-average rice cadmium concentrations[b]

| Rice cadmium concentration (μg/g) | 50-59 | | | 60-69 | | | > 69 | | |
|---|---|---|---|---|---|---|---|---|---|
| | No. | A (%) | B (%) | No. | A (%) | B (%) | No. | A (%) | B (%) |
| Control (< 0.1) | 104 | 0.96 | 0 | 80 | 2.50 | 1.25 | 64 | 0 | 0 |
| 0.19-0.29 | 477 | 0.84 | 0.42 | 377 | 2.65 | 0.80 | 268 | 13.81[c] | 4.48 |
| 0.30-0.39 | 184 | 3.26 | 2.72 | 138 | 5.07 | 0.72 | 91 | 17.58[c] | 7.69[d] |
| 0.40-0.49 | 295 | 1.69 | 0.34 | 204 | 6.86 | 2.94 | 117 | 14.53[c] | 8.55[c] |
| 0.50-0.59 | 140 | 0.71 | 0 | 120 | 10.83[d] | 5.00 | 93 | 34.41[c] | 25.81[c] |
| 0.60-0.69 | 60 | 18.33[c] | 5.00[d] | 57 | 24.56[c] | 12.28[c] | 23 | 73.91[c] | 43.48[c] |

[a] Prevalence of increased RBP in Ishikawa prefecture only

[b] From: Nogawa et al. (1978); No. = number of people examined; RBP was measured by a semiquantitative method; values in column A refer to the prevalence of RBP in urine at levels above 4 mg/litre; values in column B refer to the prevalence at values above 16 mg/litre

[c] Significant difference ($P < 0.01$) compared with control

[d] Significant difference ($P < 0.05$) compared with control

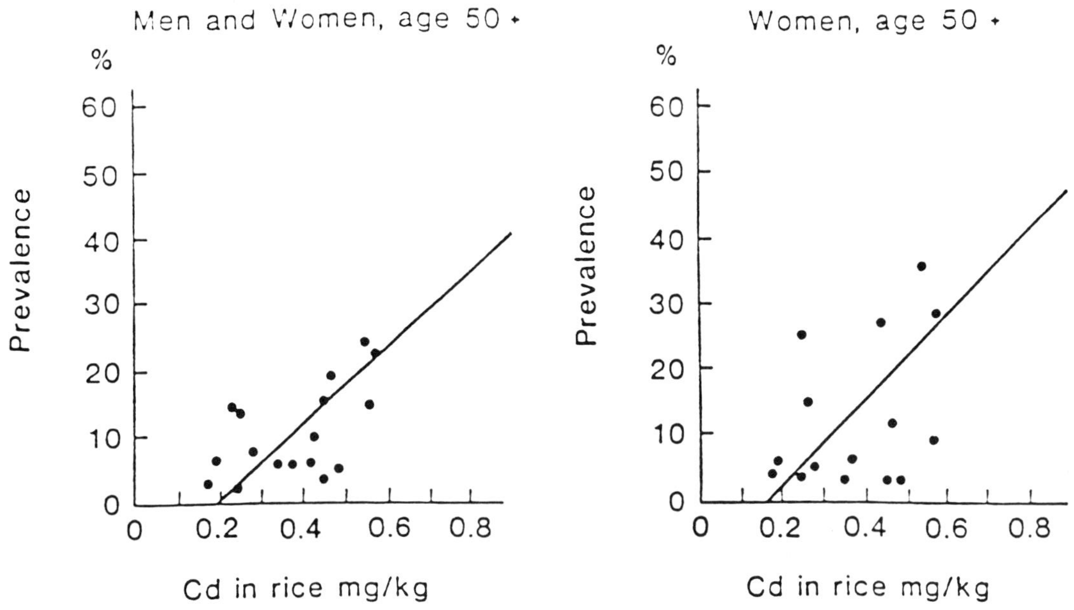


Fig. 10. Prevalence of increased urinary RBP (above 3 mg/litre) among inhabitants of villages in the Kakeashi river basin as a function of village-average cadmium concentration in rice. From: Ishikawa Prefectural Government (1976)

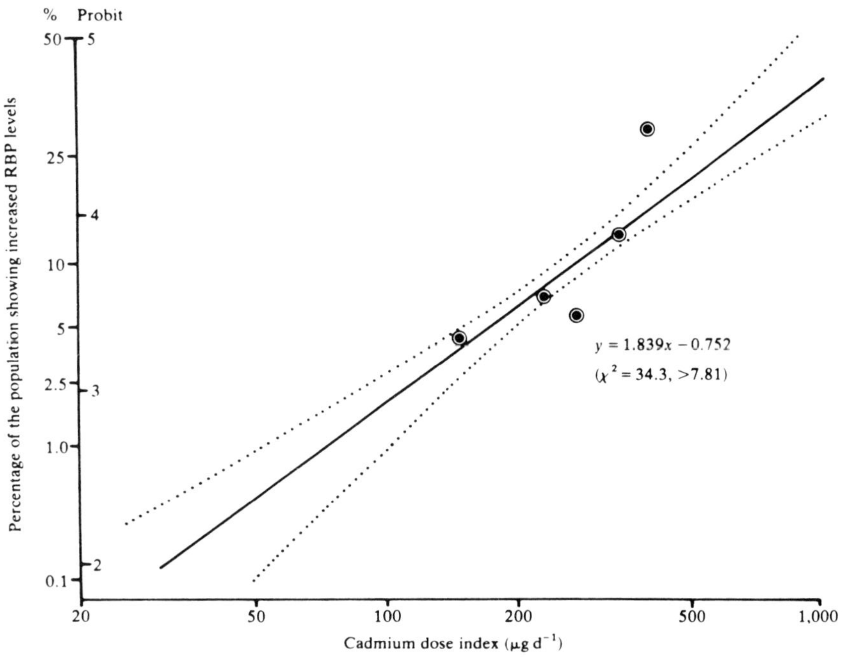


Fig. 11. Dose-response relationship for cadmium-induced increase in urinary RBP levels, with upper and lower 95% fiducial limits. From: Hutton (1983), based on data from Nogawa et al. (1978)

prevalence of tubular proteinuria varies according to age and average rice cadmium concentration. Of the 438 participants in the final detailed examinations (426 in the polluted areas and 12 in the non-polluted areas), findings of "possible proximal renal tubular dysfunction" were noted in 334 people (333 in the polluted areas and 1 in the non-polluted areas). Among these cases, 202 in the polluted areas were determined to have proximal renal tubular dysfunction and 116 of them were considered to require medical supervision in view of the severity of the dysfunction. The urinary $\beta_2$-microglobulin level in Itai-itai disease patients was on average 43 mg/litre, 100 times higher than that in the controls, and the patients investigated by Nogawa et al. (1978) had an average urinary level of $\beta_2$-microglobulin 65 times the controls.

In a later epidemiological study, Nogawa et al. (1989) investigated the dose-response relationship in 1850 cadmium-exposed and 294 non-exposed inhabitants of the Kakehashi River basin. Using a urine concentration of 1000 $\mu$g $\beta_2$-microglobulin/g creatinine as an index of renal tubular dysfunction, and the average rice cadmium concentration as an index of cadmium exposure, the authors determined linear regression equations for men and women. These are related to the prevalence of $\beta_2$-microglobulinuria and total cadmium intake and are shown in Table 19. The authors concluded that the total cadmium intake that produced an adverse effect on health was approximately 2000 mg for both men and women. On the basis of the linear regression equations shown in Table 19, an average daily cadmium intake of 440 $\mu$g/day in men and 350 $\mu$g/day in women would be expected to cause a 50% response rate (> 1000 $\mu$g $\beta_2$-microglobulin/litre urine). Response rates of 20, 10 and 5% would occur at daily intakes of 220, 150, and 110 $\mu$g/day for men and 200, 150, and 120 $\mu$g/day for women. The authors indicated that these data are in general agreement with results from other studies involving the consumption of various levels of cadmium in rice.

#### 8.3.3.5 *Akita prefecture (Kosaka area)*

Around Kosaka mine and refinery areas, increased cadmium levels in rice were first reported by the Akita Prefectural Government (1973). Kojima et al. (1976) gave further data from the Kosaka area, where the reported average cadmium level in rice varied between 0.26 and 0.56 mg/kg. The latter average was based on 41 samples where one was reported to contain 4.81 mg/kg. In the exposed area, the weighted average of cadmium levels in rice in each district according to the number of examinees was

Table 19. Linear regression equations relating total cadmium intake and prevalence of $\beta_2$-microglobulinuria[a]

| Sex | $\beta_2$-microglobulinuria | Linear regression equation[b] | Prevalence of $\beta_2$-microglobulinuria in the control group (%) | Total cadmium intake (mg)[c] |
|---|---|---|---|---|
| Male | > 1000 μg/litre | Y = 0.0076X - 10.33 | 5.3 | 2057 |
| | > 1000 μg/g creatinine | Y = 0.0083X - 7.93 | 6.0 | 1678 |
| Female | > 1000 μg/litre | Y = 0.011X - 19.61 | 3.1 | 2065 |
| | > 1000 μg/g creatinine | Y = 0.012X - 16.16 | 5.0 | 1763 |

[a] From: Nogawa et al. (1989)
[b] Y = prevalence of $\beta_2$-microglobulinuria (%); X = total cadmium intake (mg)
[c] Total cadmium intake yielding a $\beta_2$-microglobulinuria prevalence equivalent to the control group

calculated by Kojima et al. (1976) to be 0.57 mg/kg in 1973 and 0.50 mg/kg in 1974. These values, according to Kojima et al. (1976), represented the level of cadmium exposure in this study and were considered more accurate than the general average given above.

In an epidemiological investigation of the total population in the age-group 50-69 of defined geographical areas (93 out of 98 in the control area and 156 out of 190 in the exposed area participated), Kojima et al. (1976, 1977) obtained data on faecal excretion of cadmium. The cadmium level in 24-h faeces samples was analysed only for those participants who said that they defaecated once a day (64 in the control area and 118 in the polluted area). Average rates were 51 μg/day and 177 μg/day for the control and exposed groups, respectively (Kojima et al., 1976). The prevalence of proteinuria exceeding 150 mg/litre (using the Tsuchiya biuret method) and of combined proteinuria and glucosuria (test tape) was significantly higher in the exposed group than in the control group (Kojima et al., 1976).

Quantitative analysis of urinary $\beta_2$-microglobulin with radioimmunoassay (RIA) was performed on the same population (Kojima et al., 1977). The frequency distributions were log-normal, as for occupationally exposed people, and the prevalence of $\beta_2$-microglobulin excretion was 15% in the whole exposed group and 3% in the control group. The exposed and control groups showed average faecal cadmium excretion rates of 139 and 41 μg per day, respectively (Kojima et al., 1977).

The $\beta_2$-microglobulin data and cadmium intake data in the study by Kojima et al. (1977) were divided into three dose groups (Kjellström et al., 1977b) and analysed for a dose-response relationship by Hutton (1983).

Further studies of urinary $\beta_2$-microglobulin over the age range 5-90 years were reported by Saito et al. (1981). The RIA method was used, and a control area in Akita prefecture was compared with cadmium-polluted areas in Akita prefecture (Kosakai), Ishikawa prefecture (Kakehashi) (section 8.3.3.3), and Nagasaki prefecture (Tsushima) (section 8.3.3.5). For people over the age of 40, there were significant increases in average urinary $\beta_2$-microglobulin excretion in all the polluted areas. In the older age-groups, the increase was 10-100 times above control values.

In the first comprehensive study of proximal renal tubular functions performed on a population living in a cadmium-polluted area, Saito et al. (1977) conducted health examinations within Akita prefecture. Renal tubular function tests consisted of renal glucosuria, uric acid clearance, low molecular weight proteinuria, tubular reabsorption of phosphate, hydrogen carbonate threshold, acid-base balance, concentrating and acidifying ability of urine, endogenous creatinine clearance, and renal plasma flow. Of the 147 residents (97% of target population) examined, 33 (22%) had some indications of proximal renal tubular dysfunction, such as renal glucosuria and low molecular weight proteinuria. In addition, 10 subjects (7%) were diagnosed as having multiple proximal renal tubular dysfunctions. Detailed examinations revealed that none of these 10 subjects had experienced any other environmental exposures or diseases that could have caused the renal dysfunction. They were therefore diagnosed as suffering from the effects of chronic cadmium poisoning (Saito et al., 1977).

#### 8.3.3.6 *Nagasaki prefecture (Tsushima area)*

This area has been a lead and zinc mining district from ancient times (Takabatake, 1978b). Modern operations started in 1948, and mining wastes have been scattered throughout the local area. In 1952, local farmers complained about the poor growth of crops. In the 1960s, studies of cadmium pollution were carried out. Health examinations for cadmium effects have been conducted since 1966 (Takabatake, 1978b).

Early studies showed an increased prevalence of proteinuria in the most polluted village, and an average rice cadmium level of 0.75 mg/kg was reported (Takabatake, 1978b).

The study of urinary $\beta_2$-microglobulin according to age referred to in section 8.3.3.4 (Saito et al., 1981) included an exposed group of people from Tsushima. The age-specific average urinary excretions were much higher than the control values and similar to those found in the Kakehashi and Kosaka areas. For women, the Tsushima values appear higher than those of the other two polluted areas (Table 17), which may reflect the higher estimated average cadmium intake in the Tsushima area.

#### 8.3.3.7 *Other Japanese areas*

As shown in Table 7, health surveys have been performed in areas other than Fuchu, Ikuno, Kakehashi, Kosaka, and Tsushima.

According to the Japanese Cadmium Research Committee (1989), it should be emphasized, however, that no cadmium health effects, including elevated prevalence of $\beta_2$-microglobulinuria, were observed in some areas with higher levels of cadmium (daily intakes of 180-309 μg in Bandai, Fukushima; 180-380 μg in Annaka, Gunma; and 222-391 μg in Okutake river basin, Oita) even than Kosaka and Kakehashi (cadmium in rice = 0.16-0.58 mg/kg, daily intake of cadmium = 139-177 μg in Kosaka, Akita; cadmium in rice = 0.2-0.8 mg/kg, daily intake of cadmium = 160-190 μg in the Kakehashi river basin, Ishikawa). Also, no health effects were found in the Uguisuzawa river basin, Miyagi (cadmium in rice = 0.6-0.7 mg/kg), Watarase river basin, Gunma (0.32 mg/kg), Shimoda, Shizuoka (0.4-1.1 mg/kg), and Ohmuta, Fukuoka (0.72 mg/kg), even though residents consumed rice heavily contaminated with cadmium.

#### 8.3.3.8 *Belgium*

The Liege area of Belgium is known to be polluted by cadmium, mainly due to the activities of non-ferrous smelters since the end of the 19th century (Kretzschmar et al., 1980; Lauwerys et al., 1980a).

A pilot study was performed in 1979 on a group of 60 elderly non-smoking women who had spent most of their lives in the Liege area and had never been occupationally exposed to cadmium (Roels et al., 1981a). Daily intakes of cadmium ranged from 2-88 μg/day with an average of 15 μg/day. Their average blood cadmium level (1.6 μg/litre) and their urinary excretion rates of cadmium (0.093 μg/h), total protein (17.3 mg/h), amino acids (5.45 mg amino acid N/h), and albumin (1.54 mg/h) were higher than those found in a group of 70 women of the same age and socio-economic status who lived in another industrial area (Charleroi) less polluted by cadmium. Although the average excretion rate of $\beta_2$-microglobulin was greater in Liege (93.6 μg/h) than in Charleroi (22 μg/h), the difference between the geometric means was not statistically significant. The two areas were well matched with respect to their environmental pollution by sulfur dioxide, fume, suspended particles, and various metals, including lead, vanadium, nickel, chromium, and iron.

Following these observations, a mortality study and a preliminary autopsy study were undertaken (Lauwerys & De Wals, 1981; Lauwerys et al., 1984a). It was found that, although the overall mortality was not markedly different, the standard mortality ratio (SMR) and proportional mortality rate (PMR) from nephritis and nephrosis for the years 1967-1976 were higher in Liege than in Charleroi or in Belgium as a whole (SMR: Belgium, 100; Charleroi, 102; Liege, 196; PMR: Belgium, 3.3; Charleroi, 3.0; Liege, 6.0). Since the increased mortality rate for renal diseases was observed in both males and females, the influence of environmental factors other than occupation is probable.

The results of the preliminary autopsy study indicated that, in the age-group 40-60, the average body burden of cadmium was approximately twice as high in people autopsied in Liege as it was in those autopsied in a city (Brussels) less polluted by cadmium (Lauwerys et al., 1984a). The geometric mean values of cadmium concentration in the kidney cortex were 38.3 and 22.8 mg/kg wet weight in Liege and Brussels, respectively.

According to the authors of these reports, the studies performed so far in the Liege area do not refute the hypothesis that environmental pollution by cadmium in the area has increased the body burden of cadmium of the inhabitants and has affected their renal function. A large-scale morbidity and autopsy study is at present underway (Braux et al., 1987).

#### 8.3.3.9 *Shipham area in the United Kingdom*

The village of Shipham is located on the slag heaps of an old zinc mine and high levels of cadmium have been found in soil and dust (section 3.4.3).

The exposed population has been studied using both mortality and morbidity end-points. A census in 1979 identified 1092 residents, of whom 64% had resided in the village for more than 5 years, and 548 participated in a health study coordinated by the United Kingdom Department of the Environment (Barltrop & Strehlow, 1982a). A similar study of 543 age- and sex-matched individuals was performed in a nearby control village (Barltrop & Strehlow, 1982b). The daily intake of cadmium in Shipham was an average of 35 $\mu$g/day (section 4.2.4), which is about twice as high as the estimated United Kingdom national average but much lower than in polluted areas of Japan (section 5.2.4).

A health inventory was compiled by means of a questionnaire, which included information on smoking habits, alcohol consumption, medication, and occupation. Blood samples were analysed for haemoglobin, haematocrit, serum protein, $\beta_2$-microglobulin, creatinine, erythrocyte protoporphyrin, lead, and cadmium. Urine samples were analysed for total protein, creatinine, $\beta_2$-microglobulin, and cadmium.

The mean 24-h urinary concentration of cadmium for Shipham residents was 0.68 $\mu$g cadmium/g creatinine and, in the control area, 0.60 $\mu$g cadmium/g creatinine with 97.7% of values less than 3.4 $\mu$g cadmium/litre. The difference was statistically significant ($P < 0.03$) (Barltrop & Strehlow, 1982b), but the similarity of the values for average urinary cadmium concentrations between the two areas and the generally low levels of cadmium in Shipham suggest a rather low daily cadmium intake.

However, there are data showing that some individuals in Shipham had high cadmium exposures. Liver cadmium concentrations, measured by means of *in vivo* neutron activation analysis, were determine for 21 adult volunteers living in the most heavily contaminated areas of Shipham (Harvey et al., 1979). Their age range was 40-62 years (mean, 53 years) and, with one exception, they had lived in Shipham for 9-50 years (mean, 23 years). On average, half of their vegetable consumption was of local origin. The mean liver cadmium concentration was 11.0 ($\pm$ 2.0) mg/kg, compared with 2.2 ($\pm$ 2.0) mg/kg in 20 age-matched, non-Shipham controls ($P < 0.001$). The maximum concentration in the Shipham group was 28 mg/kg, which would correspond to levels in the kidney cortex of at least 200-300 mg/kg (Friberg, 1979) (section 6.4). Health effects were not studied in this investigation.

In the health study by the Department of the Environment (Barltrop & Strehlow, 1982a), the comparison of $\beta_2$-microglobulin data from the two villages showed similar distributions, and all other laboratory data, including blood pressure levels, were distributed within the normal range. However, the poor participation rate in this health study (50%), makes it difficult to interpret the findings. Another study of 31 volunteers from Shipham (Carruthers & Smith, 1979) reported a high prevalence of hypertension and LMW proteinuria, but the methodology of the study has been criticized (Hughes & Stewart, 1979; Kraemer et al., 1979).

Examination of the data in relation to soil cadmium levels showed no evidence of an increased mortality from any cause in those living in the most polluted areas. It was concluded from this study that, if cadmium contamination had any effect on the mortality pattern in Shipham, this effect was only slight and did not present a serious health hazard to residents. No case resembling Itai-itai disease has at any time been reported in Shipham. All the authors involved in the health studies in Shipham pointed to the possible protective effect of high levels of zinc also present in soil, and Kraemer et al. (1979) pointed to the need to assess dietary zinc and the intake of nutrients other than cadmium.

#### 8.3.3.10 *USSR*

Screening of populations, both environmentally and occupationally exposed, which included measurements of urinary $\beta_2$-microglobulin, has been carried out within the USSR (Likutova, 1989). Increased prevalence (up to 6%) of $\beta_2$-microglobulinuria (> 280 $\mu g/g$ creatinine) was observed in females (20-50 years old) in some of the most heavily polluted cities, i.e. Odjonikidze and Kursk. The air cadmium concentrations in these two cities were 0.085 $\mu g/m^3$ and 0.005-0.027 $\mu g/m^3$, respectively. The examination of workers (50-300 $\mu g/m^3$) exposed to cadmium also revealed an increased prevalence of $\beta_2$-microglobulinuria (up to 19% in the most heavily exposed group). The findings are in good agreement with the data presented in Table 15.

## 8.4 Conclusions

High inhalation exposure to cadmium oxide fume results in acute pneumonitis with pulmonary oedema, which may be lethal. High ingestion exposure of soluble cadmium salts causes acute gastroenteritis.

Long-term occupational exposure to cadmium has caused severe chronic effects, predominantly in the lungs and kidneys. Chronic renal effects have also been seen among the general population.

Following high occupational exposure, lung changes are primarily characterized by chronic obstructive airway disease. Early minor changes in ventilatory function tests may progress, with continued cadmium exposure, to respiratory insufficiency. An increased mortality rate from obstructive lung disease has been seen in workers with high exposure, as has occurred in the past.

The accumulation of cadmium in the renal cortex leads to renal tubular dysfunction with impaired reabsorption of, for instance, proteins, glucose, and amino acids. A characteristic sign of tubular dysfunction is an increased excretion of low molecular weight proteins in urine. In some cases, the glomerular filtration rate decreases. Increase in urine cadmium correlates with low molecular weight proteinuria and in the absence of acute exposure to cadmium may serve as an indicator of renal effect. In more severe cases there is a combination of tubular and glomerular effects, which may progress in some cases to decreased glomerular filtration. For most workers and people in the general environment, cadmium-induced proteinuria is irreversible.

Among other effects are disturbances in calcium metabolism, hypercalciuria, and formation of renal stones. High exposure to cadmium, most probably in combination with other factors such as nutritional deficiencies, may lead to the development of osteoporosis and/or osteomalacia.

There is evidence that long-term occupational exposure to cadmium may contribute to the development of cancer of the lung but observations from exposed workers have been difficult to interpret because of confounding factors. For prostatic cancer, evidence to date is inconclusive but does not support the suggestion from earlier studies of a causal relationship.

At present, there is no convincing evidence for cadmium being an etiological agent of essential hypertension. Most data speak against a blood pressure increase due to cadmium and there is no evidence of an increased mortality due to cardiovascular or cerebrovascular disease.

Data from studies on groups of occupationally exposed workers and on groups exposed in the general environment show that there is a relationship between exposure levels, exposure durations, and the prevalence of renal effects.

An increased prevalence of low molecular weight proteinuria in cadmium workers after 10-20 years of exposure to cadmium levels of about 20-50 $\mu g/m^3$ has been reported.

In polluted areas of the general environment, where the estimated cadmium intake has been 140-260 $\mu g$/day, effects in the form of increased low molecular weight proteinuria have been seen in some individuals following long-term exposure.

# 9. EVALUATION OF HUMAN HEALTH RISKS

## 9.1 Exposure assessment

### 9.1.1 General population exposure

In the ambient air, cadmium concentrations based on long-term sampling periods indicate, in most cases, a range of 0.001-0.015 $\mu g/m^3$ in rural areas, 0.005-0.05 $\mu g/m^3$ in urban areas, and up to 0.6 $\mu g/m^3$ near sources of pollution (section 5.1.1).

One cigarette usually contains 1-2 $\mu g$ cadmium, of which about 10% may be inhaled (section 5.1.3).

Among staple foods, rice and wheat usually contain less than 0.1 mg/kg and other foods usually less than 0.05 mg/kg wet weight, but liver and kidney may contain 1-2 mg/kg wet weight and certain sea-foods as much as 10 mg/kg wet weight (section 5.2). Certain animals, e.g., the horse, may accumulate considerably higher concen-trations in the liver and kidney. In polluted areas, these levels are further increased.

The content of natural waters is usually less than 1 $\mu g$/litre, but higher levels may be found near sources of pollution.

The total daily intake in non-polluted areas of most countries from food, water and air is estimated to be approximately 10-40 $\mu g$/day (food, 10-40 $\mu g$/day; water, $< 1$ $\mu g$; and air, $< 0.5$ $\mu g$/day for non-smokers).

Twenty cigarettes per day would contribute a further 2-4 $\mu g$. In polluted areas, the daily intake may be much higher, and intakes of several hundred $\mu g$/day have been reported (section 5.3.2).

### 9.1.2 Occupational exposure

Air is the main source of additional cadmium exposure for industrial workers. In many countries such exposures have now been reduced considerably. In the past, levels of several $mg/m^3$ were recorded in workplaces. Now, with proper industrial hygiene practices, levels of 0.02-0.05 $mg/m^3$ would be more typical (section 5.1.2).

### 9.1.3 Amounts absorbed from air, food, and water

The proportion of cadmium from food and water that is absorbed will depend on the chemical nature of the cadmium compounds, but estimates based on the available data indicate that gastrointestinal absorption is approximately 5%, with considerable individual variation (section 6.1.2). Similarly, the amount absorbed from the air will depend on the chemical nature and the particle size of the inhaled material. The absorption varies between 25 and 50% depending on particle size and solubility (section 6.1.1). About 10% of the cadmium inhaled in cigarette smoke is absorbed.

Thus, the average amount absorbed from food and water in a person from a non-polluted area would be about 0.5-1.3 μg/day. The absorbed amount from smoking 20 cigarettes per day would be 1-2 μg/day and that from workroom air could be many times greater (section 5.3).

## 9.2 Dose-effect relationships

### 9.2.1 Renal effects

Long-term exposure to cadmium causes renal tubular dysfunction with proteinuria, glucosuria, and aminoaciduria, as well as histopathological changes, in both experimental animals and humans (sections 7.2.1.4 and 8.2.1, respectively). These are usually the first effects to occur after ingestion or inhalation exposure. As the renal dysfunction progresses in severity, the glomeruli may also be affected and, in a few cases, the cadmium-induced damage may lead to renal failure (section 8.2.1). Daily cadmium intakes in food of 140-260 μg/day for more than 50 years or workplace air exposures of 50 $\mu g/m^3$ for more than 10 years have produced an increase in renal tubular dysfunction in some exposed populations (section 8.3.3.2).

### 9.2.2 Bone effects

Cadmium may produce bone effects in both humans and animals. The most notable clinical entity in these cases is osteomalacia, but many subjects also show osteoporosis. Animal experiments show that both can be produced by long-term cadmium exposure (section 7.2.4). In animals and humans, osteomalacia has been seen in combination with cadmium-induced renal damage. The bone effects may be linked to cadmium effects on calcium and vitamin D metabolism in the kidney. The daily intakes via food and exposure levels in air at which the bone effects occur are uncertain, but they must be higher than those causing renal effects. Bone effects have been seen among both the general population and industrial workers in the past when exposure levels were very high. Host and nutritional factors influence the development and severity of cadmium-induced bone effects.

### 9.2.3 Pulmonary effects

Chronic obstructive airway disease has been reported in a number of studies of cadmium workers (section 8.2.3). This has, in severe cases, led to an increased mortality. The dose needed to produce these effects is uncertain, but it is higher than the dose needed to produce renal effects, as most workers reported to have lung effects also had renal effects. On the other hand, many workers with renal effects, who had been exposed to cadmium oxide dust and fume, had no lung effects.

### 9.2.4 Cardiovascular effects

Some animal studies have shown that, under certain exposure conditions, increased blood pressure and effects on the myocardium occur. Studies of cadmium-exposed workers and people in the general environment have been carried out, but most data do not support the animal findings.

### 9.2.5 Cancer

There is evidence that cadmium chloride, sulfate, sulfide and oxide give rise to injection site sarcomata in the rat and that the chloride and sulfate induce interstitial cell tumours of the testis in rats and mice.

Long-term inhalation studies in rats exposed to aerosols of cadmium chloride, sulfate, and oxide fume and dust at low concentrations demonstrated a high incidence of primary lung cancer with evidence of a dose-response relationship. This has not so far been shown in other animals.

There is evidence that long-term occupational exposure to cadmium may contribute to the development of cancer of the lung, but observations from exposed workers have been difficult to interpret because of inadequate exposure data and confounding factors. The evidence to date is inconclusive, but does not support the suggestion from earlier studies that cadmium can cause prostatic cancer.

IARC (1987a) considered that there was sufficient evidence for the carcinogenicity of specified cadmium compounds in experimental animals and limited evidence for carcinogenicity in humans exposed to cadmium. A combined evaluation of human and animal data by IARC (1987b) classified cadmium as a probable human carcinogen (IARC group 2A). The IPCS Task Group found no reason to deviate from this IARC evaluation.

### 9.2.6 Critical organ and critical effect

The kidney is the critical organ for chronic cadmium poisoning. Within the kidney, the cortex is the site where the first adverse effect occurs. Therefore, in assessing dose-response relationships, the cadmium concentration in the kidney cortex is of prime importance.

The critical effect is renal tubular dysfunction, which is most often manifested as low molecular weight proteinuria. Animal studies indicate that histological changes in the renal tubules occur at a dose level lower than that needed to produce low molecular weight proteinuria.

## 9.3 Critical concentration in the kidneys

### 9.3.1 In animals

Several studies with data on both cadmium concentrations in the renal cortex and the occurrence of tubular damage were discussed in section 7.2.1. The findings were summarized in Table 12. They showed that histological tubular lesions or proteinuria was usually seen at cadmium renal cortex levels of 200-300 mg/kg wet weight. In some studies on rats, monkeys, horses, and birds, certain effects were seen at lower levels.

As no dose-response data are given in most animal studies, it may be assumed that these renal cortex levels correspond to a 50% response rate ($CC_{50}$). Naturally, the cadmium levels at which lower response rates occur would be lower.

In studies on monkeys conducted in Japan, kidney cadmium levels were related to dose and duration of exposure. At the two highest dose levels, acute liver effects occurred.

If one wishes to establish a range of values for the critical concentration in individuals at which a small but significant part of an exposed population will show effects, animal studies indicate that a renal cortex level of about 100-200 mg/kg is likely to coincide with such a range. There is some evidence that the average critical concentration ($CC_{50}$) could be as high as 300 or 400 mg/kg for the more severe signs of renal tubular damage, but such high levels should not be used as a starting point for calculations of "acceptable daily exposures".

### 9.3.2 *In humans*

Section 8.2.1.5 reviewed all available data from cases in which both renal cortex cadmium levels and renal effects were measured. Data from autopsies or biopsies have mainly been cross-sectional, i.e. the renal cadmium concentrations and the effects were measured more or less simultaneously. This has made it difficult to interpret the data from a critical concentration point-of-view, as the cases with the most severe cadmium-induced kidney dysfunction had the lowest renal cadmium levels. Cadmium is lost from the kidney when the damage progresses (section 6.5.1.2).

*In vivo* neutron activation analysis has provided a new tool for establishing the human critical concentration of cadmium in the renal cortex. Longitudinal studies measuring the renal cortex cadmium concentration several times during continued exposure can now be carried out. The cadmium level at which the first measurable signs of renal tubular dysfunction occurs can be estimated. However, only two studies using *in vivo* neutron activation have been published to date, and both of them are cross-sectional.

The renal cadmium concentrations are disproportionately low when the liver cadmium concentrations are high and renal effects have developed. Of the several methods available to estimate the average critical kidney concentration in these groups of exposed workers, the method of preference assumes that the peak for renal cortex cadmium level, plotted against liver cadmium, is equivalent to the point where renal tubular dysfunction occurs. This results in a value of 319 mg/kg tissue (based on a ratio of renal cortex cadmium to whole kidney cadmium of 1.5). There is considerable variance in the individual values, the 95% tolerance (which corresponds to a confidence interval) being in the range $\pm$ 90 mg/kg from the mean. Other studies, using similar assumptions, have reported a value of 332 mg/kg, 10% of the workers having a peak cadmium level of about 216 mg/kg tissue. A re-evaluation of the original study resulted in a calculated cadmium level of about 200 mg/kg. It was concluded that for the purposes of dose-response calculations, using a metabolic model, this concentration could be used as a starting point for renal effects occurring in an exposed population.

## 9.4 Dose-response relationships for renal effects

Two approaches can be used to estimate dose-response relationships. One employs epidemiological data from industry and the general environment studying associations between exposure and response. The other begins with a critical concentration in the kidney cortex and employs a metabolic model to calculate, on the basis of certain given assumptions, the exposure that is required to reach a critical concentration.

### 9.4.1 *Evaluation based on data on industrial workers*

Table 16 contains data from various group studies on cadmium workers. In most of these studies, the dose measurements were based on short sampling periods (hours or a few days). However, exposure may have lasted for decades, levels usually being higher in the past. The use of protective devices may also confound the picture.

As discussed in section 8.3.2, there are now several reports available that show a clear exposure-response relationship between cadmium in workplace air and the prevalence of proteinuria.

An increased prevalence of overt proteinuria, as measured by sulfosalicylic acid, trichloroacetic acid, or quantitative determination of total proteinuria, can occur after only 5-10 years of exposure to approximately 100 $\mu$g cadmium/m$^3$. If instead, the increased excretion of low molecular weight proteins (more than 97.5 percentile of control group) is used as the critical effect, 10-20% of workers would have this effect after a cumulative dose corresponding to 10-20 years of exposure to 50 $\mu$g cadmium/m$^3$. These evaluations are all based on levels of total cadmium in inhaled dust or air.

### 9.4.2 *Evaluation based on data on the general population*

As indicated in chapter 8, there exists a considerable amount of information from epidemiological studies carried out on the general population in Japan. It was shown that in some areas of high cadmium exposure the prevalence of low molecular weight proteinuria was significantly higher than in control areas. This may be considered in relation to the known cadmium concentrations in rice and the daily cadmium intakes in the affected areas (Tables 7 and 17). Contamination of drinking-water in some areas may be a complicating factor (section 8.3.3).

Taking all of the data in section 8.3.3 together, it seems that, when the most sensitive method for diagnosis of low molecular weight proteinuria is applied, there is an association between cadmium exposure and increased excretion of low molecular weight proteins among some people over 50 years of age at a daily intake of about 140-260 μg cadmium or a cumulative cadmium intake of about 2000 mg or more (for both men and women).

### 9.4.3 Evaluation based on a metabolic model and critical concentrations

Using the data on critical concentrations and kinetic models of cadmium metabolism, attempts have been made to calculate the dose-response relationship for cadmium. Assuming that cadmium in the kidney is accumulated in accordance with a one-compartment model and that a third or a quarter of the body burden of cadmium is in the kidney (and making certain other assumptions indicated in Tables 20 and 21), the daily cadmium intake via food and the occupational air concentrations needed to reach the critical concentration have been calculated (Tables 20 and 21).

As the values calculated in Tables 20 and 21 are for an average person, not all of those exposed to these levels would have reached the renal cortex cadmium concentration of 200 mg/kg tissue or their individual critical concentration. Nevertheless, these calculations produce values that are similar to the levels at which effects have been observed, and the model approach may be a useful way to quantify the response rates at levels lower than those easily measurable.

Calculations have been reported of the relationship between intake and response rates using the observed frequency distributions of daily intake and renal cortex cadmium concentrations, and utilizing multi-compartment metabolic model values in the same range as those given in Tables 20 and 21. Further development of these modelling techniques would be of value.

Table 20. Calculated daily cadmium intake via ingestion required by a non-smoker to reach a kidney cortex concentration of 200 mg/kg at age 50 (using a one-compartment model)[a]

| Gastrointestinal absorption rate (%) | Proportion of body burden in kidney | Estimated half-time in kidney cortex[b] | |
|---|---|---|---|
| | | 17 years | 30 years |
| 5 | one-third | 365 μg (286 μg) | 265 μg (208 μg) |
| 10 | | 182 μg (143 μg) | 133 μg (104 μg) |
| 5 | one-quarter | 486 μg (382 μg) | 353 μg (277 μg) |
| 10 | | 243 μg (191 μg) | 177 μg (139 μg) |

[a] The data in the table are based on the following assumptions:

gastrointestinal absorption, either 5% or 10%;
half-time in kidney cortex, either 17 years or 30 years (as reported in section 6.6.2);
one-third or one-quarter of body burden in the kidneys;
cadmium concentration in renal cortex 25% higher than renal average;
average weight of both kidneys at age 50 of 300 g for a 70-kg person or 235 g for a 55-kg person;
average cadmium concentration in foodstuffs constant during the last 50 years;
variation of daily intake with age has been disregarded since such variation would influence the values by less than 10%.

[b] Data have been calculated for a 70-kg person; values in parentheses are for a 55-kg person.

Using a single-compartment model for the accumulation of cadmium in the kidney, the average daily intake that would give rise to an average concentration of 200 mg/kg wet weight in the kidney cortex at age 50 would be 260-480 μg/day, assuming 5% gastrointestinal absorption, various biological half-times, and different proportions of the body burden in the kidneys (Table 19). Assuming a 10% absorption rate, the intake needed would be 140-260 μg per day. These estimates will vary depending on the body weight estimates for different populations.

Table 21. Calculated concentration of cadmium in industrial air required for a kidney cortex concentration of 200 mg/kg to be reached[a]

| Proportion of body burden in kidney | Estimated half-time in kidney cortex[b] | |
|---|---|---|
| | 17 years | 30 years |
| one-third | 16 μg/m$^3$ (13 μg/m$^3$) | 14 μg/m3 (11 μg/m3) |
| one-quarter | 21 μg/m$^3$ (17 μg/m$^3$) | 19 μg/m$^3$ (15 μg/m$^3$) |

[a] The data in the table are based on the following assumptions:

those assumptions given in Table 20;
exposure time of 25 years;
225 working days per year;
10 m$^3$ of air inhaled per day;
25% pulmonary absorption

[b] Data have been calculated for a 70-kg worker; values in parentheses are for an average 55-kg person

# 10. CONCLUSIONS AND RECOMMENDATIONS FOR PROTECTION OF HUMAN HEALTH

## 10.1 Conclusions

The kidney is considered the critical target organ for the general population as well as for occupationally exposed populations. Chronic obstructive airway disease is associated with long-term high-level occupational exposure by inhalation. There is some evidence that such exposure to cadmium may contribute to the development of cancer of the lung but observations from exposed workers have been difficult to interpret because of confounding factors.

### *10.1.1 General population*

Food-borne cadmium is the major source of exposure for most people. Average daily intakes from food in most areas not polluted with cadmium are 10-40 μg. In polluted areas the value has been found to be several hundred μg per day. In non-polluted areas, uptake from heavy smoking may equal cadmium intake from food.

An association between cadmium exposure and increased urinary excretion of low molecular weight proteins has been noted in humans with a life-long daily intake of about 140-260 μg cadmium, or a cumulative intake of about 2000 mg or more.

### *10.1.2 Occupationally exposed population*

Occupational exposure to cadmium is mainly by inhalation but includes additional intakes through food and tobacco. The total cadmium level in air varies according to industrial hygiene practices and type of workplace. There is an exposure-response relationship between airborne cadmium levels and proteinuria. An increase in the prevalence of low molecular weight proteinuria may occur in workers after 10-20 years of exposure to cadmium levels of about 20-50 $\mu g/m^3$. *In vivo* measurement of cadmium in the liver and kidneys of people with different levels of cadmium exposure have shown that about 10% of workers with a kidney cortex level of 200 mg/kg and about 50% of people with a kidney cortex level of 300 mg/kg would have renal tubular proteinuria.

## 10.2 Recommendations for protection of human health

a) Measures to increase recycling of cadmium should be systematically examined and promising ideas encouraged.

b) Information on the importance of minimizing waste discharge of cadmium, particularly into surface waters, should be supplied to countries.

c) Public health measures for protection from cadmium exposures would be improved by:

   i) collection of more data from countries on cadmium levels in foodstuffs and the environment;

   ii) determination of tissue cadmium levels and monitoring of health parameters in non-exposed populations and in those living near mines or smelters or exposed to elevated levels of the metal in foodstuffs;

   iii) technical assistance to developing countries for the training of staff, particularly for cadmium analysis;

   iv) development of means of reducing cadmium exposure by, for instance, improved working conditions and the dissemination of information on the proper use of fertilizers (which sometimes contain high levels of cadmium), techniques for the disposal of cadmium-containing wastes, etc.

# 11. FURTHER RESEARCH

a) There is a need for improved analytical techniques for measuring cadmium species and biological indicators of cadmium exposure/toxicity, such as $\beta_2$-microglobulin, in various matrices, and for international centres for quality assurance and training.

b) The assessment of human exposure to cadmium from all media needs to be improved by increased monitoring of cadmium levels in the environment. Changes in cadmium levels with time are of particular importance.

c) Populations with $\beta_2$-microglobinuria (both those in the workplace and in the general environment) should be longitudinally investigated to determine the nature, severity, and prognosis of adverse health effects associated with this finding. Further research is needed on $\beta_2$-microglobulin as a biological indicator of exposure and effect.

d) International collaborative efforts should be encouraged to examine further the role of cadmium in the development of human cancer. Both the general population and industrial workers should be studied with special emphasis on the development of a common format for analysing and presenting data and the collection of additional information on exposure to cadmium, tobacco, and other confounding factors. Multiple exposures must be considered. It is proposed that a collaborative study coordinated by an international body (e.g., IARC) should include the existing cohorts in order to obtain better exposure data. It should also collect both exposure and effects data in a standardized manner, so that the results of different studies may be more readily compared. A further approach would be to perform a collaborative prospective study identifying all those workers who have shown evidence of an effect of cadmium on the kidney and who would therefore be considered to have had unusually heavy exposure. In such a study, both morbidity and mortality data would be collected. Outcome would be studied not only for cancer but also for sequelae to renal dysfunction.

e) Existing occupational cohorts should be linked, where possible, to regional cancer registers to determine the incidence of prostatic cancer (morbidity) in relation to cadmium exposure.

f) To understand the mechanism(s) of cancer induction, experimental studies on the bioavailability of cadmium at the target site and the interactions between zinc and cadmium would be of value. The role of metallothionein induction in the target cells of the respiratory tract and its relationship to such phenomena as DNA damage and repair and oncogene protein structure would be of interest.

g) Further information on the long-term health consequences of cadmium exposures in the general environment is essential, with emphasis on renal dysfunction and other end-points such as neurotoxicity and immunotoxicity.

h) Studies of the effects of cadmium on calcium-phosphorus metabolism and bone density should be conducted on female workers to clarify whether these workers are at special risk in the occupational setting. The effect of cadmium on the placenta and subsequent effects on the fetus, especially in multiple pregnancies, need further study.

i) The effects of various nutritional deficiencies and of exposure to other metals on the transport, accumulation, and toxicity of cadmium should be investigated with special reference to bone toxicity. These studies should be conducted in humans and experimental animals with respect to age, sex, dose-dependence, biological half-time, and estimation of critical concentration.

j) To provide additional scientific support for the assessment of human health risks from cadmium exposure, studies in experimental animals addressing the following issues should be initiated:

- mechanism of cadmium transport into the cell and factors controlling the process;

- mechanism of cadmium-induced toxicity with particular emphasis on kidney and bone and the role(s) of non-metallothionein-bound cadmium in these processes;

- mechanisms of cadmium-induced calcuria and the relationship of this phenomenon to tubular proteinuria and osteomalacia.

# 12. PREVIOUS EVALUATIONS BY INTERNATIONAL BODIES

The carcinogenic potential of cadmium was evaluated in 1976 by the International Agency for Research on Cancer (IARC, 1976) and re-evaluated in 1987 (IARC, 1987a). It was concluded in the re-evaluation that there was limited evidence that cadmium and cadmium compounds are carcinogenic in humans. Sufficient evidence was available to show that cadmium and specified cadmium compounds cause cancer in experimental animals. Cadmium was classified as a probable human carcinogen (group 2A) (IARC, 1987a,b).

To prevent adverse pulmonary and renal effects the following health-based limits for occupational exposure to cadmium fumes and respirable dust were proposed by WHO (WHO, 1980): 250 $\mu$g Cd/m$^3$ for short-term exposures provided the recommended time-weighted average (40 h/week) of 10 $\mu$g Cd/m$^3$ is respected. It was further recommended that control measures be applied when cadmium levels in urine and blood of individuals exceed 5 $\mu$g Cd/g creatinine and 5 $\mu$g Cd/litre of whole blood, respectively.

A drinking-water guideline value of 0.005 mg/litre has been set for cadmium by the World Health Organization (WHO, 1984).

Cadmium was evaluated by a WHO Working Group developing air quality guidelines (WHO, 1987). Based on non-carcinogenic effects, the following recommendations were made:

a) in rural areas, levels of < 1-5 ng/m$^3$ should not be allowed to increase, and

b) in urban and industrialized areas without agricultural activities, levels of 10-20 ng/m$^3$ may be tolerable. However, increases in the present levels of airborne cadmium should not be permitted (WHO, 1987).

At the thirty-third meeting of the Joint FAO/WHO Expert Committee on Food Additives and Food Contaminants, the previous recommendation was reaffirmed, i.e. the provisional tolerable weekly cadmium intake of 400-500 $\mu$g for an adult should not be exceeded (WHO, 1989).

Regulatory standards established by national bodies in several countries and the EEC are summarized in the legal file of the International Register of Potentially Toxic Chemicals (IRPTC, 1987).

## REFERENCES


ADAMS, R.G. (1980) Osteopathy associated with tubular nephropathy in employees in an alkaline battery factory. In: Shigematsu, I. & Nomiyama, K., ed. Cadmium-induced osteopathy, Tokyo, Japan Public Health Association, pp. 67-73.

ADAMS, R.C., HARRISON, J.G., & SCOTT, P. (1969) The development of cadmium-induced proteinuria, impaired renal function, and osteomalacia in alkaline battery workers. Q. J. Med., **38**: 425-443.

ADAMSSON, E. (1979) Long-term sampling of air-borne cadmium dust in an alkaline battery factory. Scand. J. Work Environ. Health, **5**: 178-187.

ADES A.E. & KAZANTZIS, G. (1988) Lung cancer in a non-ferrous smelter: the role of cadmium. Br. J. ind. Med., **45**: 435-442.

AHLGREN, L. & MATTSON, S. (1981) Cadmium in man measured *in vivo* by X-ray fluorescence analysis. Phys. Med. Biol., **26**: 19-26.

AHLMARK, S., AXELSSON, B., FRIBERG, L., & PISCATOR, M. (1961) Further investigations into kidney function and proteinuria in chronic cadmium poisoning. Int. Congr. occup. Health, **13**: 201-203.

AHOKAS, R.A. & DILTS, P.V. (1979) Cadmium uptake by the rat embryo as a function of gestational age. Am. J. Obstet. Gynecol., **135**: 219-222.

AKITA PREFECTURAL GOVERNMENT (1973) [Surveys on environmental pollution due to heavy metals: Annual report of the Technical Centre for Environmental Protection], Akita, Japan, Akita Prefectural Government (Akita Prefecture Report No. 1: 1970 and 1971) (in Japanese).

AL-HADDAD, I.K., CHETTLE, D.R., FLETCHER, J.G., & FREMLIN, J.H. (1981) A transportable system for measurement of kidney cadmium *in vivo*. Int. J. appl. Radiat. Isot., **32**: 109-112.

ALLANSON, M. & DEANESLY, R. (1962) Observations on cadmium damage and repair in rat testes and the effects on the pituitary gonadotrophs. J. Endocrinol., **24**: 453-462.

ALLOWAY, B.J., THORNTON, I., SMART, G.A., SHERLOCK, J.C., & QUINN, M.J. (1988) Metal availability. Sci. total Environ., **75**: 41-69.

ALT, F. (1981) [Comparative determination of cadmium in blood by four different methods.] Fresenius Z. anal. Chem., **308**: 137-142 (in German).

ANDERSEN, A. (1979) [Lead, cadmium, copper and zinc in the Danish diet], Copenhagen, Statens Levendsmiddelinstitut, 89 pp, (Report No. 40) (in Danish).

ANDERSEN, O., HAGERSTRAND, I., & NORDBERG, G.F. (1982) Effect of the chelating agent sodium tripolyphosphate on cadmium toxicity in mice. Environ. Res., **29**: 54-61.

ANDERSEN, O., RONNE, M., & NORDBERG, G.F. (1983) Effects of inorganic metal salts on chromosome length in human lymphocytes. Hereditas, **98**: 65-70.

ANDERSON, C. & DANYLCHUK, K.D. (1979) Plasma levels of immunoreactive parathyroid hormone in dogs chronically exposed to low levels of cadmium chloride. J. environ. Pathol. Toxicol., **2**: 1151-1159.

ANDERSSON, A. (1977) Heavy metals in Swedish soils: on their retention, distribution, and amounts. Swed. J. agric. Res., **7**: 7-20.

ANDERSSON, A. & HAHLIN, M. (1981) Cadmium effects from phosphorus fertilization in field experiments. Swed. J. agric. Res., **11**: 3-10.

ANDERSSON, K., ELINDER, C.-G., HOGSTEDT, C., KJELLSTROM, T., & SPANG, G. (1984) Mortality among cadmium workers in a Swedish battery factory. In: Proceedings of the 4th International Cadmium Conference, Munich, March 1983, London, Cadmium Association, pp. 152-154.

ANDO, M., SHIMIZU, M., SAYATA, Y., TANIMURA, A., & TOBE, M. (1981) The inhibition of vitamin D-stimulated intestinal calcium transport in rats after continuous oral administration of cadmium. Toxicol. appl. Pharmacol., **61**: 297-301.

ANKE, M. & SCHNEIDER, H.J. (1974) [The trace element concentration of human kidneys as a function of age and sex.] Z. Urol., **67**: 357-362 (in German).

ANKE, M., HENNIG, A., SCHNEIDER, H.J., GROPPEL, B., GRUN, M., PARTSCHEFELD, M., & LUDKE, H. (1976) [The influence of the toxic element cadmium on the metabolism and health of animals and human.] Math. naturwiss. Unterr. **R 25**: 241-261 (in German).

AOKI, A. & HOFFER, A.P. (1978) Re-examination of the lesions in rat testis caused by cadmium. Biol. Reprod., **18**: 579-591.

APOSTOLOV, CH. (1979) [The effects on the upper respiratory tract from long-term work with cadmium.] In: [Proceedings of the Cadmium Symposium, Jena, 1-3 August, 1977], Jena, Friedrich-Schiller Universität, pp. 322-325 (in German).

ARMSTRONG, B.G. & KAZANTZIS, G. (1983) The mortality of cadmium workers. Lancet, **1**: 1424-1427.

ARMSTRONG, B.G. & KAZANTZIS, G. (1985) Prostatic cancer and chronic respiratory and renal disease in British cadmium workers: a case control study. Br. J. ind. Med., **42**: 540-545.

ASOKAN, P., DIXIT, R., MUKHTAR, H., & MURTI, C.R.K. (1981) Effect of cadmium on hepatic mixed function oxidases during the early development of rats; possible protective role of metallothionein. Biochem. Pharmacol., **30**(22): 3095-3098.

AUFDERHEIDE, M., MOHR, U., THIEDEMANN, K.-U., & HEINRICH, U. (1990) Quantification of hyperplastic areas in hamster lungs after chronic inhalation of different cadmium compounds. Toxicol. environ. Chem., **27**: 173-180.

AXELSSON, B. (1963) Urinary calculus in long-term exposure to cadmium. In: Sangro, P., Akoun, G., & Beate, H.L., ed. Proceedings of the 14th International Congress on Occupational Health, Madrid, Amsterdam, Excerpta Medica Foundation, pp 939-942 (International Congress Series No. 62).

AXELSSON, B. & PISCATOR, M. (1966a) Renal damage after prolonged exposure to cadmium: an experimental study. Arch. environ. Health, **12**: 360-373.

AXELSSON, B. & PISCATOR, M. (1966b) Serum proteins in cadmium-poisoned rabbits with special reference to haemolytic anaemia. Arch. environ. Health, **12**: 374-381.

AXELSSON, B., DAHLGREN, S.E., & PISCATOR, M. (1968) Renal lesions in the rabbit after long-term exposure to cadmium. Arch. environ. Health, **17**: 24-28.

BAADER, E.W. (1951) [Health concerns and occupational medicine: chronic cadmium poisoning.] Dtsch. med. Wschr., **76**: 484-487 (in German).

BAETZ, R.A. & KENNER, C.T. (1974) Determination of low levels of cadmium in foods using a chelating ion exchange resin. J. Assoc. Off. Anal. Chem., **57**: 14-17.

BARLOW, S.M. & SULLIVAN, F.M. (1982) Cadmium and its compounds. In: Barlow, S.M., ed. Reproductive hazards of industrial chemicals. An evaluation of animal and human data, New York, London, Academic Press, pp. 137-173.

BARLTROP, D. & STREHLOW, C.D. (1982a) Clinical and biochemical indices of cadmium exposure in the population of Shipham. In: Proceedings of the 3rd International Cadmium Conference, Miami, Florida, 3-5 February, 1981, London, Cadmium Association, pp. 112-113.

BARLTROP, D. & STREHLOW, C.D. (1982b) Cadmium and health in Shipham. Lancet, **2**: 1394-1395.

BARR, M., Jr (1973) The teratogenicity of cadmium chloride in two stocks of Wistar rats. Teratology, **7**: 237-242.

BARRETT, H.M. & CARD, B.Y. (1947) Studies on the toxicity of inhaled cadmium. II. The acute lethal dose of cadmium oxide for man. J. ind. Hyg. Toxicol., **29**: 286-293.

BARRETT, H.M., IRWIN, D.A., & SEMMONS, E. (1947) Studies on the toxicity of inhaled cadmium. I. The acute toxicity of cadmium oxide by inhalation. J. ind. Hyg. Toxicol., 29: 279-285.

BARTHELEMY, P. & MOLINE, R. (1946) Intoxication chronique par l'hydrate de cadmium, son signe précoce: La bague jaune dentaire. Paris Med., **1**: 71-78.

BAUCHINGER, M., SCHMID, E., EINBRODT, H.J., & DRESP, J. (1976) Chromosome aberrations in lymphocytes after occupational exposure to lead and cadmium. Mutat. Res., **40**: 57-62.

BECKMAN, I., MOVITZ, I., NYGREN, M., & SLORACH, S. (1979) [Release of lead and cadmium from glazed and enamelled foodware.] Var Föda, **31**: 193-197 (in Swedish).

BEHNE, D. (1980) Problems of sampling and sample preparation for trace element analysis in the health sciences. In: Bratter, P. & Schramel, P., ed. Trace element analytical chemistry in medicine and biology. Proceedings of the First International Workshop, Neuherberg, Federal Republic of Germany, Berlin, New York, Walter de Gruyter & Co., pp. 769-782.

BERLIN, M. & FRIBERG, L. (1960) Bone marrow activity and erythrocyte destruction in chronic cadmium poisoning. Arch. environ. Health, **1**: 478-486.

BERMAN, E. (1967) Determination of Cd, Tl, and Hg in biological materials by AA. At. Absorpt. Newsl., **6**: 57-60.

BERNARD, A.M., ROELS, H., HUBERMONT, G., BUCHET, J.P., MASSON, P.L., & LAUWERYS, R. (1976) Characterization of the proteinuria in cadmium-exposed workers. Int. Arch. occup. environ. Health, **38**: 19-30.

BERNARD, A.M., BUCHET, J.P., ROELS, H., MASSON, P., & LAUWERYS, R. (1979) Renal excretion of proteins and enzymes in workers exposed to cadmium. Eur. J. clin. Invest., **9**: 11-22.

BERNARD, A.M., LAUWERYS, R., & GENGOUX, P. (1981) Characterization of the proteinuria induced by prolonged oral administration of cadmium in female rats. Toxicology, **20**: 345-357.

BERNARD, A.M., MOREAU, D., & LAUWERYS, R. (1982a) Latex immunoassay of retinol binding protein. Clin. Chem., **28**: 1167-1171.

BERNARD, A.M., MOREAU, D., & LAUWERYS, R. (1982b) Comparison of retinol-binding protein and β-2-microglobulin determination in urine for the early detection of tubular proteinuria. Clin. Chim. Acta, **126**: 1-9.

BETON, D.C., ANDREWS, G.S., DAVIES, H.J., HOWELLS, L., & SMITH, G.F. (1966) Acute cadmium fume poisoning: five cases with one death from renal necrosis. Br. J. ind. Med., **23**: 292-301.

BHATTACHARYYA, M.H., WHELTON, B.D., PETERSON, D.P., CARNES, B.A., MORETTI, E.S., TOOMEY, J.M., & WILLIAMS, L.L. (1988) Skeletal changes in multiparous mice fed a nutrient-sufficient diet containing cadmium. Toxicology, **50**: 193-204.

BIGGIN, H.C., CHEN, N.S., ETTINGER, K.V., FREMLIN, J.H., MORGAN, W.D., NOWOTNY, R., CHAMBERLAIN, M.J., & HARVEY, T.C. (1974) Cadmium by *in vivo* neutron activation analysis. J. radioanal. Chem., **19**: 207-214.

BJORNBERG, A. (1963) Reactions to light in yellow tattoos from cadmium sulfide. Arch. Dermatol., **88**: 267-271.

BLEJER, H.P. (1966) Death due to cadmium oxide fumes. Ind. Med. Surg., **35**: 363-364.

BOISSET, M. & BOUDENE, C. (1981) Effect of a single exposure to cadmium oxide fumes on rat lung microsomal enzymes. Toxicol. appl. Pharmacol., **57**: 335-345.

BONNELL, J.A. (1955) Emphysema and proteinuria in men casting copper-cadmium alloys. Br. J. ind. Med., **12**: 181-197.

BONNELL, J.A., KAZANTZIS, G., & KING, E. (1959) A follow-up study of men exposed to cadmium oxide fume. Br. J. ind. Med., **16**: 135-147.

BONNELL, J.A., ROSS, J.H., & KING, E. (1960) Renal lesions in experimental cadmium poisoning. Br. J. ind. Med., **17**: 69-80.

BOULEY, G., DUBREUIL, A., DESPAUX, N., & BOUDENE, C. (1977) Toxic effects of cadmium microparticles on the respiratory system. An experimental study on rats and mice. Scand J. Work Environ. Health, **3**: 116-121.

BOYLE, E.A., SCALTER, F., & EDMOND, J.M. (1976) On the marine geochemistry of cadmium. Nature (Lond.), **263**: 42-44.

BOZELKA, B.E. & BURKHOLDER, P.M. (1982) Inhibition of mixed leukocyte culture response in cadmium-treated mice. Environ. Res., **27**: 421-432.

BRADY, F.O. (1982) The physiological function of metallothionein. Trends biochem. Sci., **7**: 143-145.

BRAUX, P., CLAEYS-THOREAU, F., DE PLAAN, P., DUCOFFRE, G., AMERY, A., LAUWERYS, R., RONDIA, D., STAESSEN, J., ROELS, H., SARTON, F., BERNARD, A., & BUCHET, J.P. (1987) The Belgian Cadmibel Study: Environmental exposure to cadmium and lead-health effects. In: Lindberg, S.F. & Hutchinson, J.C., ed. Proceedings of an International Conference on Heavy Metals in the Environment, New Orleans, Edinburgh, CEP Consultants Ltd, pp. 349-351.

BRYAN, G.W., LANGSTON, W.J., & HUMMERSTONE L.G. (1980) The use of biological indicators of heavy-metal contamination in estuaries with special reference to an assessment of the biological availability of metals in estuarine sediments from south-west Britain, Citadel Hill, Devon, The Marine Biological Association of the United Kingdom, 73 pp (Occasional Publication No. 1).

BUATMENARD, P. & ARNOLD, M. (1978) Heavy-metal chemistry of atmospheric particulate matter emitted by Mount Etna volcano. Geogr. Res. Lett., **5**: 245-248.

BUCHAUER, M.J. (1972) The effects of zinc and cadmium pollution on forest vegetation. Bull. New Jersey Acad. Sci., **17**(2): 42.

BUCHET, J.P., ROELS, H., BERNARD, A., & LAUWERYS, R. (1980) Assessment of renal function of workers exposed to inorganic lead, cadmium or mercury vapor. J. occup. Med., **22**: 741-748.

BUCHET, J.P., ROELS, H., LAUWERYS, R., VANDEVOORDE, A., & PYCKE, J.M. (1983) Oral daily intake of cadmium, lead, manganese, copper, chromium, mercury, calcium, zinc, and arsenic in Belgium: a duplicate meal study. Food chem. Toxicol., **21**: 19-24.

BUCKE, D., NORTON, M.G., & ROLFE, M.S. (1983) Field assessment of effects of dumping wastes at sea: II. Epidermal lesions and abnormalities of fish in the outer Thames estuary, London, Ministry of Agriculture, Fisheries and Food, 29 pp (Technical Report No. 72).

BUI, T.-H-. LINDSTEN, J., & NORDBERG, G.F. (1975) Chromosome analysis of lymphocytes from cadmium workers and Itai-itai patients. Environ. Res., **9**: 187-195.

BUS, J.S., VINEGAR, A., & BROOKS, S.M. (1978) Biochemical and physiologic changes in lungs of rats exposed to a cadmium chloride aerosol. Am. Rev. respir. Dis., **118**: 573-580.

BUTLER, E.A. & FLYNN, F.V. (1958) The proteinuria of renal tubular disorders. Lancet, **2**: 978-980.

BUXTON, R., St J. (1956) Respiratory function in cadmium alloy casters. Part 2. The estimation of the total lung volume, its subdivisions and the mixing coefficient. Br. J. ind. Med., **13**: 36-40.

CARLIER, Y., COLLE, A., TACHON, P., BOUT, D., & CAPRON, A. (1981) Quantification of B-2-microglobulin by enzyme inhibition immunoassay. J. immunol. Methods, **40**: 231-238.

CARLSON, L.A. & FRIBERG, L. (1957) The distribution of cadmium in blood after repeated exposure. Scand. J. clin. lab. Invest., **9**: 1-4.

CARLTON-SMITH, C.H. (1987) Effects of metals in sludge-treated soils on crops, Medmenham, United Kingdom, Water Research Centre (Environment TR/251).

CARROLL, R.E. (1966) The relationship of cadmium in the air to cardiovascular disease death rates. J. Am. Med. Assoc., **198**: 267-269.

CARRUTHERS, M. & SMITH, B. (1979) Evidence of cadmium toxicity in a population living in a zinc-mining area. Lancet, **1**: 845-847.

CARVALHO, F.M., TAVARES, T.M., SILVANY-NETO, A.M., LIMA, M.E.C., & ALT, F. (1986) Cadmium concentrations in blood of children living near a lead smelter in Bahia, Brazil. Environ. Res., **40**: 437-449.

CASTERLINE, J.L. & YIP, G. (1975) The distribution and binding of cadmium in oyster, soybean, rat liver, and kidney. Arch. environ. Contam. Toxicol., **3**: 319-329.

CASTILHO, P.D. & HERBER, R.F.M. (1977) The rapid determination of cadmium, lead, copper, and zinc in whole blood by atomic absorption spectrometry with electrothermal atomization. Improvements in precision with a peak-shape monitoring device. Anal. Chim. Acta, **94**: 269-270.

CHADWICK, M.J. & LINDMAN, N. (1982) Environmental implications of expanded coal utilization, Oxford, New York, Pergamon Press, 283 pp.

CHAPATWALA, K.D., HOBSON, M., DESAIAH, D., & RAJANNA, B. (1982) Effect of cadmium on hepatic and renal gluconeogenic enzymes in female rats. Toxicol. Lett., **12**: 27-34.

CHAUBE, S., NISHIMURA, H., & SWINYARD, C.A. (1973) Zinc and cadmium in normal human embryos and fetuses. Arch. environ. Health, **26**: 237-240.

CHERIAN, M.G. (1980) The synthesis of metallothionein and cellular adaptation to metal toxicity in primary kidney epithelial cell cultures. Toxicology, **17**: 225-231.

CHERIAN, M.G. & GOYER, R.A. (1978) Metallothioneins and their role in the metabolism and toxicity of metals. Life Sci., **23**: 1-10.

CHERIAN, M.G. & NORDBERG, M. (1983) Cellular adaptation in metal toxicology and metallothionein. Toxicology, **28**: 1-15.

CHERIAN, M.G. & SHAIKH, Z.A. (1975) Metabolism of intravenously-injected cadmium-binding protein. Biochem. biophys. Res. Commun., **65**: 863-869.

CHERIAN, M.G. & VOSTAL, J.J. (1977) Biliary excretion of cadmium in rat. I. Dose-dependent biliary excretion and the form of cadmium in the bile. J. Toxicol. environ. Health, **2**: 945-954.

CHERIAN, M.G., GOYER, R.A., & RICHARDSON, L.D. (1976) Cadmium metallothionein-induced nephropathy. Toxicol. appl. Pharmacol., **38**: 399-408.

CHERIAN, M.G., GOYER, R.A., & VALBERG, L.S. (1978) Gastrointestinal absorption and organ distribution of oral cadmium chloride and cadmium metallothionein in mice. J. Toxicol. environ. Health, **4**: 861-868.

CHERNOFF, N. (1973) Teratogenic effects of cadmium in rats. Teratology, **8**: 29-32.

CHERRY, W.H. (1981) Distribution of cadmium in human tissues. In: Nriagu, J.O., ed. Cadmium in the environment - II, New York, Chichester, John Wiley & Sons, pp. 111-122.

CHIQUOINE, A.D. (1965) Effect of cadmium chloride on the pregnant albino mouse. J. Reprod. Fertil., **10**: 263-265.

CHRISTOFFERSON, J.O. & MATTSON, S. (1983) Polarized X-rays in XRF-analysis for improved *in vivo* detectability of cadmium in man. Phys. Med. Biol., **28**: 1135-1144.

CHUMBLEY, C.G. & UNWIN, R.J. (1982) Cadmium and lead content of vegetable crops grown on land with a history of sewage sludge application. Environ. Pollut. Ser. B Chem. Phys., **4**(3): 231-238.

CHUNG, J., NARTY, N.O., & CHERIAN, M.G. (1986) Metallothionein levels in liver and kidney of Canadians - A potential indicator of environmental exposure to cadmium. Arch. environ. Health, **41**(5): 319-323.

CIRKT, M. & TICHY, M. (1974) Excretion of cadmium through bile and intestinal wall in rats. Br. J. ind. Med., **31**: 134-139.

CLARKSON, T.W. & KENCH, J.E. (1956) Urinary excretion of amino acids by men absorbing heavy metals. Biochem. J., **62**: 361-372.

CNATTINGIUS, S., AXELSSON, O., EKLUND, G., & LINDMARK, G. (1985) Smoking, maternal age and fetal growth. Obstet. Gynecol., **66**: 449-452.

COLE, G.M. & BAER, L.S. (1944) Food poisoning from cadmium. US nav. med. Bull., **43**: 398-399.

COMMISSION OF THE EUROPEAN COMMUNITIES (1978) Criteria (dose-effect relationships) for cadmium, Oxford, Pergamon Press, 202 pp (Report of the Working Group of Experts prepared for the CEC, Directorate-General for Social Affairs, Health and Safety Directorate, Luxembourg).

COOPERATIVE RESEARCH COMMITTEE ON ITAI-ITAI DISEASE (1967) [Report on a study on the so-called Itai-itai disease], Tokyo, Ministry of Education and Ministry of Health and Welfare, 206 pp (in Japanese).

COPIUS-PEEREBOOM, J.W., DE VOOGT, P., VAN HATTUM, B., VAN DER VELDE, W., & COPIUS-PEEREBOOM-STEGEMAN, J.H.J. (1979) The use of the human placenta as a biological indicator for cadmium exposure. In: International Conference on Management and Control of Heavy Metals in the Environment, London, September, 1979, Edinburgh, CEP Consultants Ltd, pp. 8-10.

COUSINS, R.J., BARBER, A.K., & TROUT, J.R. (1973) Cadmium toxicity in growing swine. J. Nutr., **103**: 964-972.

CULBARD, E.B., THORNTON, I., WATT, J., WHEATLEY, M., MOORCROFT, S., & THOMPSON, M. (1988) Metal contamination in British urban dusts and soils. J. environ. Qual., **17**(2): 227-234.

CVETKOVA, R.P. (1970) [Materials on the study of the influence of cadmium compounds on the generative function.] Gig. Tr. prof. Zabol., **14**: 31-33 (in Russian with English summary).

DALHAMN, T. & FRIBERG, L. (1954) The effect of cadmium on blood pressure and respiration and the use of dimercaprol (BAL) as antidote. Acta pharmacol. toxicol., **10**: 199-203.

DANIELSSON, L.-G., JAGNER, D., JOSEFSON, M., & WESTERLUND, S. (1981) Computerized potentiometric stripping analysis for the determination of cadmium, lead, copper, and zinc in biological materials. Anal. Chim. Acta, **127**: 147-156.

DASTON, G.P. (1982) Fetal zinc deficiency on a mechanism for cadmium-induced toxicity to the developing rat lung and pulmonary surfactant. Toxicology, **24**: 55-63.

DAVIES, N.T., BREMNER, I., & MILLS, C.F. (1973) Studies on the induction of a low molecular weight zinc-binding protein in rat liver. Biochem. Soc. Trans., **1**: 985-988.

DAVIS, R.D. (1984) Cadmium - a complex environmental problem. Part II. Cadmium in sludges used as fertilizer. Experientia (Basel), **40**(2): 117-126.

DAVIS, R.D. & COKER, E.G. (1980) Cadmium in agriculture, with special reference to the utilisation of sewage sludge on land, Medmenham, United Kingdom, Water Research Centre (Technical Report TR/139).

DAVISON, A.G., NEWMAN-TAYLOR, A.J., DARBYSHIRE, J., & CHETTLE, D.R. (1988) Cadmium fume inhalation and emphysema. Lancet, **1**: 663-667.

DEAVEN, L.L. & CAMPBELL, E.W. (1980) Factors affecting the induction of chromosomal aberrations by cadmium in Chinese hamster cells. Cytogenet. cell Genet., **26**: 251-260.

DECKER, L.E., BYERRUM, R.U., DECKER, C.F., HEPPERT, C.A., & LANGHAM, R.F. (1958) Chronic toxicity studies. I. Cadmium administered in drinking-water to rats. Am. Med. Assoc. Arch. ind. Health, **18**: 228-231.

DEKNUDT, G. & DEMINATTI, M.L. (1978) Chromosome studies in human lymphocytes after *in vitro* exposure to metal salts. Toxicology, **10**: 67-75.

DEKNUDT, G. & LEONARD, A. (1975) Cytogenetic investigations on leukocytes of workers from a cadmium plant. Environ. Physiol. Biochem., **5**: 319-327.

DELVES, H.T. (1970) Micro-sampling method for the rapid determination of lead in blood by atomic absorption spectrophotometry. Analyst, **95**: 431-438.

DELVES, H.T. (1982) Analytical techniques for measuring cadmium in blood. In: Proceedings of the International Workshop on Biological Indicators on Cadmium Exposure Diagnostic and Analytical Reliability, Luxembourg, 7-9 July, 1982, Luxembourg, Commission of the European Communities, pp. 1-15.

DELVES, H.T. & WOODWARD, J. (1981) Determination of low levels of cadmium in blood by electrothermal atomization and atomic absorption spectrophotometry. At. Spectrosci., **2**: 65-67.

DENCKER, L., DANIELSSON, B., KHAYAT, A., & LINDGREN, A. (1983) Disposition of metals in the embryo and fetus. In: Clarkson, T.W., Nordberg, G.F., & Sager, P.R., ed. Reproductive and developmental toxicity of metals, New York, London, Plenum Press, pp. 607-632.

DERVAN, P.A. & HAYES, J.A. (1979) Peribronchiolar fibrosis following acute experimental lung damage by cadmium aerosol. J. Pathol., **128**: 143-148.

DE SILVA, P.E. & DONNAN, M.B. (1981) Chronic cadmium poisoning in a pigment-manufacturing plant. Br. J. ind. Med., **38**: 76-86.

DUCE, R.A., HOFFMAN, G.L., & ZELLER, W.H. (1975) Atmospheric trace metals at remote northern and southern hemisphere sites: pollution or natural(?). Science, **187**: 59-61.

DUCE, R.A., ARIMOTO, R., RAY, B.J., UNNI, C.K., & HARDER, P.J. (1983) Atmospheric trace elements at Enewetok Atoll. I. Concentration, sources, and temporal variability. J. geophys. Res., **88**: 5321-5342.

DUDLEY, R.E., SVOBODA, D.J., & KLAASSEN, C.O. (1982) Acute exposure to cadmium causes severe liver injury in rats. Toxicol. appl. Pharmacol., **65**: 302-313.

DUDLEY, R.E., GAMMAL, L.M., & KLAASSEN, C.D. (1985) Cadmium-induced hepatic and renal injury in chronically exposed rats: likely role of hepatic cadmium-metallothionein in nephrotoxicity. Toxicol. appl. Pharmacol., **77**: 414-426.

DUGGAN, M.J., INSKIP, M.J., RUNDLE, S.A., & MOORCROFT, J.S. (1985) Lead in playground dust and on the hands of schoolchildren. Sci. total Environ., **44**(1): 65-79.

DUNPHY, B. (1967) Acute occupational cadmium poisoning: a critical review of the literature. J. occup. Med., **9**: 22-26.

EDIGER, R.D. & COLEMAN, R.L. (1973) Determination of cadmium in blood by a Delves cup technique. At. Absorpt. Newsl., **12**: 3-6.

EDLING, C., ELINDER, C.G., & RANDMA, E. (1986) Lung function in workers using cadmium containing solders. Br. J. ind. Med., **43**: 657-662.

ELINDER, C.-G. & PANNONE, M. (1979) Biliary excretion of cadmium. Environ. Health Perspect., **28**: 123-126.

ELINDER, C.-G. & PISCATOR, M. (1978) Cadmium and zinc in horses. In: Kirchgessner, M., ed. Proceedings of the 3rd International Conference on Trace Elements in Man and Animals, Freising-Weihenstephan, Federal Republic of Germany, 25-29 July, 1977, Freising-Weihenstephan, Munich Technical University, pp. 569-572.

ELINDER, C.-G., KJELLSTRÖM, T., LIND, B., & LINNMAN, L. (1976) Cadmium concentration in kidney cortex, liver, and pancreas among autopsied Swedes. Arch. environ. Health, **31**: 292-302.

ELINDER, C.-G., KJELLSTRÖM, T., LINNMAN, L., & PERSHAGEN, G. (1978) Urinary excretion of cadmium and zinc among persons from Sweden. Environ. Res., **15**: 473-484.

ELINDER, C.-G., JONSSON, L., PISCATOR, M., & RAHNSTER, B. (1981a) Histopathological changes in relation to cadmium concentration in horse kidneys. Environ. Res., **26**: 1-21.

ELINDER, C.-G., NORDBERG, M., PALM, B., & PISCATOR, M. (1981b) Cadmium, zinc, and copper in horse kidney and in horse liver metallothionein: comparison with kidney cortex. Environ. Res., **26**: 22-32.

ELINDER, C.-G., KJELLSTRÖM, T., LIND, B., LINNMAN, L., PISCATOR, M., & SUNDSTEDT, K. (1983) Cadmium exposure from smoking cigarettes: variations with time and country where purchased. Environ. Res., **32**: 220-227.

ELINDER, C.-G., EDLING, C., LINDBERG, E., KAGEDAL, B., & VESTERBERG, O. (1985a) Assessment of renal function in workers previously exposed to cadmium. Br. J. ind. Med., **42**: 754-760.

ELINDER, C.-G., EDLING, C., LINDBERG, E., KAGEDAL, B., & VESTERBERG, O. (1985b) B-2-microglobulinuria among workers previously exposed to cadmium: follow-up and dose-response analyses. Am. J. ind. Med., **8**: 553-564.

ELINDER, C.-G., KJELLSTRÖM, T., HOGSTEDT, C., ANDERSON, K., & SPENG, G. (1985c) Cancer mortality of cadmium workers. Br. J. ind. Med., **42**: 651-655.

ELLIS, K.J., MORGAN, W.D., ZANZI, I., YASUMURA, S., VARTSKY, D.D., & COHN, S.H. (1981a) Critical concentrations of cadmium in human renal cortex: dose-effect studies in cadmium smelter workers. J. Toxicol. environ. Health, **7**: 691-703.

ELLIS, K.J., YASAMURA, S., & COHN, S.H. (1981b) Hair cadmium content: is it a biological indicator of the body burden of cadmium for the occupationally-exposed worker? Am. J. ind. Med., **2**: 323-333.

ELLIS, K.J., YUEN, K., YASAMURA, S., & COHN, S.H. (1984) Dose-response analysis of cadmium in man: body burden vs kidney dysfunction. Environ. Res., **33**: 216-226.

ELLIS, K.J., COHN, S.H., & SMITH, T.J. (1985) Cadmium inhalation exposure estimates: their significance with respect to kidney and liver cadmium burden. J. Toxicol. environ. Health, **15**: 173-187.

ENGSTROM, B. (1979) Factors influencing metabolism and toxicity of cadmium. Experimental studies on mice with special reference to chelating agents, Stockholm, Department of Environmental Hygiene, Karolinska Institute, 80 pp (Doctoral thesis).

ENGSTROM, B. & NORDBERG, G.F. (1978) Effects of detergent formula chelating agents on the metabolism and toxicity of cadmium in mice. Acta pharmacol. toxicol., **43**: 387-397.

ENGSTROM, B. & NORDBERG, G.F. (1979a) Dose dependence of gastrointestinal absorption and biological half-time of cadmium in mice. Toxicology, **13**: 215-222.

ENGSTROM, B. & NORDBERG, G.F. (1979b) Factors influencing absorption and retention of oral $^{109}Cd$ in mice: age, pretreatment, and subsequent pretreatment with non-radioactive cadmium. Acta pharmacol. toxicol., **45**: 315-324.

ENGVALL, J. & PERK, J. (1985) Prevalence of hypertension among cadmium-exposed workers. Arch. environ. Health, **40**: 185-190.

ESSING, H.-G., SCHALLER, K.-H., SZADOWSKI, D., & LEHNERT, G. (1969) [Normal cadmium load on man from food and beverages.] Arch. Hyg. Bakteriol., **153**: 490-494 (in German).

EVRIN, P.E., PETERSON, P.A., WIDE, L., & BERGGARD, I. (1971) Radioimmunoassay of $\beta$-2-microglobulin in human biological fluids. Scand. J. clin. lab. Invest., **28**: 439-443.

EYBL, V. & SYKORA, J. (1966) [The protection by chelating agents in acute cadmium poisoning.] Acta biol. med. Germ., **16**: 61-64 (in German).

FELDMAN, S.L. & COUSINS, R.G. (1973) Influence of cadmium on the metabolism of 25-hydroxycholecalciferol in chicks. Nutr. Rep. Int., **8**: 251-259.

FELTEN, T.L. (1979) A preliminary report of cadmium-induced chromosomal changes in somatic and germinal tissues of C57BI/6J male mice. Genetics, **88**: 26-27.

FERM, V.H. (1971) Developmental malformations induced by cadmium. Biol. Neonat., **19**: 101-107.

FERM, V.H. (1972) The teratogenic effects of metals on mammalian embryos. Adv. Teratol., **5**: 51-75.

FERM, V.H. & CARPENTER, S.J. (1968) The relationship of cadmium and zinc in experimental mammalian teratogenesis. Lab. Invest., **18**: 429-432.

FERNANDEZ, F.J. & KAHN, H.L. (1971) The determination of lead in whole blood by atomic absorption spectrophotometry with the "Delves Sampling Cup" technique. At. Absorpt. Newsl., **10**: 1-5.

FINGERLE, H., FISCHER, G., & CLASSEN, H.G. (1982) Failure to produce hypertension in rats by chronic exposure to cadmium. Food chem. Toxicol., **20**: 301-306.

FITZHUGH, O.G. & MEILLER, F.H. (1941) The chronic toxicity of cadmium. J. Pharmacol. exp. Ther., **72**: 15-20.

FLANAGAN, P.R., MCLELLAN, J.S., HAIST, J., CHERIAN, M.G., CHAMBERLAIN, M.J., & VALBERG, L.S. (1978) Increased dietary cadmium absorption in mice and human subjects with iron deficiency. Gastroenterology, **74**: 841-846.

FLETCHER, J.C., CHETTLE, D.R., & AL-HADDAD, I.K. (1982) Experience with the use of cadmium measurements of liver and kidney. J. radioanal. Chem., **71**: 547-560.

FORSTNER, U. (1980) Cadmium in the environment, Part I. In: Nriagu, J.O., ed. Cadmium in polluted sediments, New York, Chichester, John Wiley & Sons, pp. 305-363.

FOULKES, E.C. (1982) Role of metallothionein in transport of heavy metals. In: Foulkes, E.C., ed. Biological roles of metallothionein, Amsterdam, Oxford, New York, Elsevier Science Publishers, pp. 131-140.

FOWLER, B.A. & NORDBERG, G.F. (1978) The renal toxicity of cadmium metallothionein: morphometric and X-ray microanalytical studies. Toxicol. appl. Pharmacol., **46**: 609-623.

FOWLER, B.A., JONES, H.S., BROWN, H.W., & HASEMAN, J.K. (1975) The morphological effects of chronic cadmium administration on the renal vasculature of rats given low and normal calcium diets. Toxicol. appl. Pharmacol., **34**: 233-252.

FOX, M.R.S. & FRY, B.E., Jr (1970) Cadmium toxicity decreased by dietary ascorbic acid supplements. Science, **169**: 989-991.

FOX, M.R.S., FRY, B.E., Jr, HARLAND, B.F., SCHERTAL, M.E., & WEEKS, C.E. (1971) Effect of ascorbic acid on cadmium toxicity in the young coturnix. J. Nutr., **101**: 1295-1305.

FRANCAVILLA, S., MOSCARDELLI, S., FRANCAVILLA, F., CASASANTA, N., PROPERZI, G., MARTINI, M., & SANTIEMMA, V. (1981) Acute cadmium intoxication: influence of cyproterone acetate on the testis and epididymis of the rat. Arch. Andrology, **6**: 1-11.

FRANK, A., PETERSON, L., & MORNER, T. (1981) [Lead and cadmium levels in organs from moose, deer, and hare.] Svensk Veterinaertidn., **33**: 151-156 (in Swedish).

FRIBERG, L. (1948a) Proteinuria and kidney injury among workmen exposed to cadmium and nickel dust. J. ind. Hyg. Toxicol., **30**: 32-36.

FRIBERG, L. (1948b) Proteinuria and emphysema among workers exposed to cadmium and nickel dust in a storage battery plant. Proc. Int. Congr. Ind. Med., **9**: 641-644.

FRIBERG, L. (1950) Health hazards in the manufacture of alkaline accumulators with special reference to chronic cadmium poisoning. Acta med. Scand., **138**(Suppl. 240): 1-124.

FRIBERG, L. (1952) Further investigations on chronic cadmium poisoning. A study on rabbits with radioactive cadmium. Arch. ind. occup. Med., **5**: 30-36.

FRIBERG, L. (1955) Iron and liver administration in chronic cadmium poisoning and studies on the distribution and excretion of cadmium. Experimental investigations in rabbits. Acta pharmacol., **11**: 168-178.

FRIBERG, L. (1979) Cadmium in Shipham. Lancet, **1**: 823-824.

FRIBERG, L. (1988) Biological monitoring of toxic metals, New York, London, Plenum Press, pp. 103-126.

FRIBERG, L. & NYSTROM, A. (1952) [Aspects on the prognosis of chronic cadmium poisoning.] Läkartidningen, **49**: 2629-2639 (in Swedish).

FRIBERG, L. & VAHTER, M. (1983) Assessment of exposure to lead and cadmium through biological monitoring: results of a UNEP/WHO global study. Environ. Res., **30**: 95-128.

FRIBERG, L., PISCATOR, M., NORDBERG, G., & KJELLSTRÖM, T. (1974) Cadmium in the environment, 2nd ed., Cleveland, Ohio, CRC Press, 248 pp.

FRIBERG, L., KJELLSTRÖM, T., NORDBERG, G., & PISCATOR, M. (1975) Cadmium in the environment. III. A toxicological and epidemiological appraisal, Washington, DC, US Environmental Protection Agency, Office of Research and Development, 217 pp.

FRIBERG, L., ELINDER, C.G., KJELLSTRÖM, T., & NORDBERG, G.F. (1985) Cadmium and health, a toxicological and epidemiological appraisal. Vol. I. Exposure, dose and metabolism, Cleveland, Ohio, CRC Press, 209 pp.

FRIBERG, L., ELINDER, C.G., KJELLSTRÖM, T., & NORDBERG, G.F. (1986) Cadmium and health, a toxicological and epidemiological appraisal. Vol. II. Effects and response, Cleveland, Ohio, CRC Press, 303 pp.

FUKUHARA, M., BOULEY, G., GODIN, J., GIRARD, F., BOISSET, M., & BOUDENE, C. (1981) Effects of short-term inhalation of cadmium oxides on rabbit pulmonary microsomal enzymes. Biochem. Pharmacol., **30**: 715-720.

FUKUSHIMA, I. (1978) Environmental pollution and health effects - Gunma Prefecture. In: Tsuchiya, K., ed. Cadmium studies in Japan: a review, Amsterdam, Oxford, New York, Elsevier Science Publishers, pp. 170-181.

FUKUSHIMA, M., ISHIZAKI, A., SAKAMOTO, M., & KOBAYASHI, E. (1973) [Cadmium concentrations in rice eaten by farmers in the Jinzu river basin.] Jpn J. Hyg., **28**: 406-415 (in Japanese with English summary).

FUKUSHIMA, M., ISHIZAKI, A., NOGAWA, K., SAKAMOTO, M., & KOBAYASHI, E. (1974) [Epidemiological studies on renal failure of inhabitants in "Itai-itai" disease endemic district. I. Some urinary findings of inhabitants living in and around the endemic district of the Jinzu river basin.] Jpn. J. Public Health, **21**: 65-73 (in Japanese with English summary).

FUKUSHIMA, I., NAKAJIMA, T., HIGUSHI, Y., TSUCHIYA, T., NAKAJIMA, K., TANAKA, T., FUKAMACHI, M., ARAI, K., & SHIONO, T. (1975) [Results of screening examination on tubular dysfunction in aged persons confined to bed], Tokyo, Japan Public Health Association, pp. 138-142 (Kankyo Hoken Report No. 36) (in Japanese).

FUKUYAMA, Y. (1972) [Proteinuria in Itai-itai disease and cadmium poisoning.] Med. Biol., **84**: 41-46 (in Japanese).

FULLMER, C.S., OKU, T., & WASSERMAN, R.H. (1980) Effect of cadmium administration on intestinal calcium absorption and vitamin D-dependent calcium-binding protein. Environ. Res., **22**: 386-399.

GABBIANI, G. (1966) Action of cadmium chloride on sensory ganglia. Experientia (Basel), **22**: 261-264.

GABBIANI, G., BAIC, D., & DELZIEL, C. (1967) Toxicity of cadmium for the central nervous system. Exp. Neurol., **18**: 154-160.

GABBIANI, G., BADONNEL, M.-C., MATHEWSON, S.M., & RYAN, G.G. (1974) Acute cadmium intoxication: early selective lesions of endothelial clefts. Lab. Invest., **30**: 686-695.

GALE, T.F. (1973) The interaction of mercury with cadmium and zinc in mammalian embryonic development. Environ. Res., **6**: 95-105.

GARDINER, P.E., OTTAWAY, J.M., & FELL, G.S. (1979) Accuracy of the direct determination of cadmium in urine by carbon-furnace atomic absorption spectrometry. Talanta, **26**: 841-847.

GARTY, M., WONG, K.-L., & KLAASSEN, C.D. (1981) Redistribution of cadmium to blood of rats. Toxicol. appl. Pharmacol., **59**: 548-554.

GERVAIS, J. & DELPECH, P. (1963) L'intoxication cadmique. Arch. Mal. prof. Méd. Trav. Séc. soc., **24**: 803-815.

GESAMP (1984) IMO/FAO/UNESCO/WMO/IAEA/UN/UNEP Joint Group of Experts on the Scientific Aspects of Marine Pollution: Report of the Fourteenth Session, Vienna, 26-30 March, 1984, Vienna, International Atomic Energy Agency (Reports and Studies No. 21).

GESAMP (1985) IMO/FAO/UNESCO/WMO/IAEA/UN/UNEP Joint Group of Experts on the Scientific Aspects of Marine Pollution: Atmospheric transport of contaminants into the mediterranean region, Athens, Geneva, World Meteorological Association (Reports and Studies No. 26).

GESAMP (1987) IMO/FAO/UNESCO/WMO/WHO/IAEA/UN/UNEP Joint Group of Experts on the Scientific Aspects of Marine Pollution: Report of the seventeenth session, Rome, Geneva, World Health Organization (Reports and Studies No. 31).

GLEIT, C.E. (1965) High-frequency electrodeless discharge system for ashing organic matter. Anal. Chem., **37**: 314-315.

GLENNAS, A. & RUGSTAD, H.E. (1984) Cadmium uptake and metabolism in cultured cells. Environ. Health Perspect., **54**: 45-50.

GOERING, P.L., SQUIBB, K.S., & FOWLER, B.A. (1985) Calcuria and proteinuria during cadmium-metallothionein-induced proximal tubule cell injury. In: Memphill, D., ed. Proceedings of the University of Missouri's 19th Annual Conference on Trace Substances in Environmental Health, 3-6 June 1985, Columbia, Missouri, University of Missouri Press, pp. 22-35.

GOYER, R.A., CHERIAN, M.G., & DELAQUERRIERE-RICHARDSON, L. (1984) Correlation of parameters of cadmium exposure with onset of cadmium-induced nephropathy in rats. J. environ. Pathol. Toxicol. Oncol., **5**: 89-100.

GOYER, R.A., HAUST, M.D., & CHERIAN, M.G. (1989) Localization of metallothionein in placenta. Toxicologist, **9**: 130.

GREENBERG, R.R., GORDON, G.E., ZOLLER, W.H., JACKO, R.B., NEVENDORF, D.W., & YOST, K.J. (1978) Composition of particles emitted from the Nocosia municipal incinerator. Environ. Sci. Technol., **12**(12): 1329-1332.

GUNN, S.A. & GOULD, T.C. (1957) Selective accumulation of $^{115}$Cd by cortex of rat kidney. Proc. Soc. Exp. Biol. Med., **96**: 820-823.

GUNN, S.A. & GOULD, T.C. (1970) Cadmium and other mineral elements. In: Johnson, A.D., Gomes, W.R., & VanDeMark, N.L., ed. The testis, New York, London, Academic Press, Vol. 3, pp. 377-481.

GUNN, S.A., GOULD, T.C., & ANDERSON, W.A.D. (1961) Zinc protection against cadmium injury to rat testis. Arch. Pathol., **71**: 274.

GUNN, S.A., GOULD, T.C., & ANDERSON, W.A.D. (1963) Cadmium-induced interstitial cell tumours in rats and mice and their prevention by zinc. J. Natl Cancer Inst., **31**: 745-759.

GUNN, S.A., GOULD, T.C., & ANDERSON, W.A.D. (1965) Strain differences in susceptibility of mice and rats to cadmium-induced testicular damage. J. Reprod. Fertil., **10**: 273-275.

GUNN, S.A., GOULD, T.C., & ANDERSON, W.A.D. (1967) Specific response of mesenchymal tissue to cancerogenesis by cadmium. Arch. Pathol., **83**: 439-499.

GUNN, S.A., GOULD, T.C., & ANDERSON, W.A.D. (1968) Specificity in protection against lethality and testicular toxicity from cadmium. Proc. Soc. Exp. Biol. Med., **128**: 591-595.

GUTHRIE, B.E. & ROBINSON, M.F. (1977) Daily intakes of manganese, copper, zinc, and cadmium by New Zealand women. Br. J. Nutr., **38**: 55-63.

GUTHRIE, J. (1964) Histological effects of intratesticular injections of cadmium chloride in domestic fowl. Br. J. Cancer, **18**: 255-260.

HADDOW, A., ROE, F.J.C., DUKES, C.E., & MITCHELEY, B.C.V. (1964) Cadmium neoplasia: sarcomata at the site for injection of cadmium sulfate in rats and mice. Br. J. Cancer, **18**: 667-673.

HADLEY, J.G., CONKLIN, A.W., & SANDERS, C.L. (1979) Systemic toxicity of inhaled cadmium oxide. Toxicol. Lett., **4**: 107-111.

HAGA, Y. & YAMAWAKI, T. (1974) [Research report on environmental health studies of heavy metal pollution], Akita, Japan, Akita Prefecture Institute of Public Health (Report No. 18) (in Japanese).

HAGINO, N. & YOSHIOKA, K. (1961) [A study on the etiology of Itai-itai disease.] J. Jpn. Orthop. Assoc., **35**: 812-815 (in Japanese).

HAHN, R., EWERS, O., JERMANN, E., FREIER, E., BROCKHAUS, A., & SCHLIPKOTER, H.W. (1987) Cadmium in kidney cortex of inhabitants of North-West Germany: in relationship to age, sex, smoking and environmental pollution by cadmium. Int. Arch. occup. environ. Health, **59**: 165-176.

HALVORSEN, C. & STEINNES, E. (1975) Simple and precise determination of Zn and Cd in human liver by neutron activation analysis. Z. anal. Chem., **274**: 199-202.

HAMILTON, D.L. & VALBERG, L.S. (1974) Relationship between cadmium and iron absorption. Am. J. Physiol., **227**: 1033-1037.

HAMMER, D.I., FINKLEA, J.F., HENDRICKS, R.H., SHY, C.M., & HORTON, R.J.M. (1971) Hair trace metal levels and environmental exposure. Am. J. Epidemiol., **93**: 84-92.

HAMMER, D.I., FINKLEA, J.F., CREASON, J.P., SANDIFER, S.H., KEIL, J.E., PRIESTER, L.E., & STARA, J.F. (1972) Cadmium exposure and human health effects. In: Hemphill, D.D., ed. Trace substances in environmental health - V, Columbia, Missouri, University of Missouri Press, pp. 269-283.

HAMMER, D.I., COLUCCI, A.V., HASSELBLAD, V., WILLIAMS, M.E., & PINKERTON, C. (1973) Cadmium and lead in autopsy tissues. J. occup. Med., **15**: 956-963.

HARADA, A. (1987) [Results of fifteen years health examinations on cadmium workers in a cadmium pigment factory (2nd report)], Tokyo, Japan Public Health Association, pp. 219-233 (Kankyo Hoken Report No. 53) (in Japanese).

HARDY, H.L. & SKINNER, J.B. (1947) The possibility of chronic cadmium poisoning. J. ind. Hyg. Toxicol., **29**: 321-325.

HARRISON, H.E., BUNTING, H., ORDWAY, N., & ALBRINK, W.S. (1947) The effects and treatment of inhalation of cadmium chloride aerosols in the dog. J. ind. Hyg. Toxicol., **29**: 302-314.

HARRISON, J.F., LUNT, G.S., SCOTT, P., & BLAINEY, J.D. (1968) Urinary lysozyme, ribonuclease, and low molecular weight protein in renal disease. Lancet, **1**: 371-375.

HARVEY, T.C., MCLELLAN, J.S., THOMAS, B.J., & FREMLIN, J.H. (1975) Measurement of liver cadmium concentrations in patients and industrial workers by neutron activation analysis. Lancet, **1**: 1269-1272.

HARVEY, T.C., CHETTLE, D.R., FREMLIN, J.H., AL-HADDAD, I.K., & DOWNES, S.P.M. (1979) Cadmium in Shipham. Lancet, **1**(8115).

HASSELMANN, J., KOENIG, W., RICHTER, F.W., STEINER, U., & WATJEN, U. (1977) Application of PIXE to trace element analysis in biological tissues. Nucl. Instrum. Methods, **142**: 163-169.

HASSLER, E., LIND, B., & PISCATOR, M. (1983) Cadmium in blood and urine related to present and past exposure. A study of workers in an alkaline battery factory. Br. J. ind. Med., **40**: 420-425.

HAYES, J.A., SNIDER, G.L., & PALMER, K.C. (1976) The evolution of biochemical damage in the rat lung after acute cadmium exposure. Am. Rev. respir. Dis., **113**: 121-130.

HEATH, J.C. & WEBB, M. (1967) Content and intracellular distribution of the inducing metal in the primary rhabdomyosarcomata induced in the rat by cobalt, nickel, and cadmium. Br. J. Cancer, **21**: 768-779.

HEATH, J.C., DANIEL, M.R., DINGLE, J.T., & WEBB, M. (1962) Cadmium as a carcinogen. Nature (Lond.), **193**: 592-593.

HEINDRYCKX, R., DEMUYNCK, M., DAMS, R., JANSSENS, M., & RAHN, K.A. (1974) Mercury and cadmium in Belgian aerosols. In: Problems of the contamination of man and his environment by mercury and cadmium. CEC European Colloquium, July, 1973, Luxembourg, Commission of the European Communities, pp. 135-148.

HEINRICH, U., PETERS, L., ERNST, H., RITTINGHAUSEN, S., DASENBROCK, C., & KÖNIG, H. (1989) Investigation on the carcinogenic effects of various cadmium compounds after inhalation exposure in hamsters and mice. Exp. Pathol., **37**: 253-258.

HENDERSON, R.F., REBAR, A.H., PICKRELL, J.A., & NEWTON, G.J. (1979) Early damage indicators in the lung III. Biochemical and cytological response of the lung to inhaled metal salts. Toxicol. appl. Pharmacol., **50**: 123-136.

HENKE, G., SACHS, H.W., & BOHN, G. (1970) [Cadmium determination by neutron activation analysis of liver and kidneys from children and young people.] Arch. Toxikol., **26**: 8-16 (in German).

HICKEY, R.J., SCHOFF, E.P., & CLELLAND, R.C. (1967) Relationship between air pollution and certain chronic disease death rates. Arch. environ. Health, **15**: 728-738.

HIRATA, H. (1981) Annaka: land polluted mainly by fumes and dust from a zinc smelter. In: Kitagishi, K. & Yamane, I., ed. Heavy metal pollution in soils of Japan, Tokyo, Japan Scientific Societies Press, pp. 149-163.

HISLOP, J.S. (1980) Choice of analytical method. In: Bratter, P. & Schramel, P., ed. Trace element analytical chemistry in medicine and biology, Berlin, New York, Walter de Gruyter Co., pp. 747-767.

HOFFMAN, E.O., COOK, J.A., DILUZIO, N.R., & COOVER, J.A. (1975) The effects of acute cadmium administration in the liver and kidney of the rat. Lab. Invest., **32**: 655-661.

HOFFMAN, L., PUTZKE, H.-P., KAMPEHL, H.-J., RUSSBULT, R., GASE, P., SIMONN, C., ERDMANN, T., & HUCKSTORF, C. (1985) Carcinogenic effects of cadmium on the prostate of the rat. J. Cancer Res. clin. Oncol., **109**: 193-199.

HOFFMAN, L., PUTZKE, H.-P., BENDEL, L., ERDMANN, T., & HUCKSTORF, C. (1988) Electron microscopic results on the ventral prostate of the rat after $CdCl_2$ administration. A contribution towards etiology of the cancer of the prostate. J. Cancer Res. clin. Oncol., **114**: 273-278.

HOLDEN, H. (1969) Cadmium toxicology. Lancet, **2**: 57.

HOLDEN, H. (1980a) A mortality study of workers exposed to cadmium fumes. In: Proceedings of the 2nd International Cadmium Conference, Cannes, 6-8 February, 1979, London, Cadmium Association, pp. 211-216.

HOLDEN, H. (1980b) Health status of European cadmium workers. In: Proceedings of the Seminar on Occupational Exposure to Cadmium, London, 20 March, 1980, London, Cadmium Association, pp. 33-37.

HOLMBERG, R.E., Jr & FERM, V.H. (1969) Inter-relationship of selenium, cadmium, and arsenic in mammalian teratogenesis. Arch. environ. Health, **18**: 873-877.

HORSTOWA, H., SIKORSKI, M., & TYBORSKI, H. (1966) [Chronic cadmium poisoning in the clinical and radiological picture.] Med. Prac. occup. Med., **17**: 13-25 (in Polish).

HOVMAND, M.F., TJELL, J.C., & MOSBACK, H. (1983) Plant uptake of airborne cadmium. Environ. Pollut., **30**: 27-38.

HUEL, G., BOUDENE, C., & IBRAHIM, M.A. (1981) Cadmium and lead content of maternal and newborn hair: relationship to parity, birth weight and hypertension. Arch. environ. Health, **36**: 221-227.

HUGHES, E.G. & STEWART, M. (1979) Cadmium in Shipham. Lancet, **1**: 973-974.

HURLEY, L.S., GOWAN, J., & SWENERTON, H. (1971) Teratogenic effects of short-term and transitory zinc deficiency in rats. Teratology, **4**: 199-204.

HUTCHISON, F., WAI, C.M., JERNEGAN, M., HICKS, W., & LONG, R. (1979) Distribution of cadmium, lead and zinc in soil around two phosphate processing plants in Pocatello, Idaho. Bull. environ. Contam. Toxicol., **22**: 426-429.

HUTTON, M., ed. (1982) Cadmium in the European Community: a prospective assessment of sources, human exposure, and environmental impact, London, Monitoring and Assessment Research Centre, Chelsea College, University of London, 100 pp (MARC Report Number 26).

HUTTON, M. (1983) Evaluation of the relationships between cadmium exposure and indicators of kidney function, London, Monitoring and Assessment Research Centre, Chelsea College, University of London, 46 pp (MARC Report Number 29).

HUTTON, M. & SYMON, C. (1986) The quantities of cadmium, lead, mercury and arsenic entering the U.K. environment from human activities. Sci. total Environ., **57**: 129-150.

HUTTON, M., WADGE, A., & MILLIGAN, P.J. (1988) Environmental levels of cadmium and lead in the vicinity of a major refuse incinerator. Atmos. Environ., **22**: 411-416.

HYOGO PREFECTURAL GOVERNMENT (1972) [Results of a comprehensive study on cadmium pollution in the area around Ikuno Mine and an outline of counter-measures], Kobe, Japan, Hyogo Prefectural Government, 156 pp (in Japanese).

IARC (1976) Cadmium, nickel, some epoxides, miscellaneous industrial chemicals, and general considerations on volatile anaesthetics, Lyon, International Agency for Research on Cancer, 306 pp (Monographs on the Evaluation of Carcinogenic Risk of Chemicals to Man, Vol. 11).

IARC (1987a) Genetic and related effects: an updating of selected IARC Monographs from Volumes 1 to 42. Lyon, International Agency for Research on Cancer (IARC Monographs on the Evaluations of Carcinogenic Risks to Humans, Supplement 6).

IARC (1987b) Cadmium and cadmium compounds. In: Overall evaluations of carcinogenicity: an updating of IARC Monographs, Volumes 1 to 42, Lyon, International

Agency for Research on Cancer, pp. 139-142 (IARC Monographs on the Evaluation of the Carcinogenic Risk of Chemicals to Humans, Supplement 7).

ILZSG (1988) International Lead and Zinc Study Group - World directory: Lead and zinc mines and primary metallurgical works, London, Chameleon Press, Ltd.

INSKIP, H., BERAL, V., & MCDOWALL, M. (1982) Mortality in Shipman residents: 40-year follow-up. Lancet, **1**: 896-899.

IRPTC (1987) IRPTC legal file 1986, Geneva, International Register of Potentially Toxic Chemicals, United Nations Environment Programme, Vol. I.

ISHIKAWA PREFECTURAL GOVERNMENT (1976) [Report on the health examination of residents of the Kakehashi river basin], Kanazawa, Japan, Health and Welfare Department, 104 pp (in Japanese).

ISHIZAKI, A. (1969) [On the so-called Itai-itai disease.] J. Jpn. Med. Soc., **62**: 242-249 (in Japanese).

ISHIZAKI, A. (1972) [Concentrations of cadmium and zinc in the organs of Itai-itai disease patients and residents in the Hokuriku region], Tokyo, Japanese Society of Public Health, pp. 154-157 (Kankyo Hoken Report No. 11) (in Japanese).

ISHIZAKI, A., FUKUSHIMA, M., SAKOMOTO, M., KURACHI, T., & HAYASHI, E. (1969) [Cadmium content of rice eaten in the Itai-itai disease area.] In: [Proceedings of the 27th Annual Meeting of the Japanese Society of Public Health], Tokyo, Japan Public Health Association, p. 111 (in Japanese).

ISHIZAKI, A., FUKUSHIMA, M., & SAKAMOTO, M. (1971) [Contents of cadmium and zinc in organs of Itai-itai disease patients and residents of Hakuriku district.] Jpn. J. Hyg., **26**: 268-273 (in Japanese).

ISHIZU, S., MINAMI, M., SUZUKI, A., YAMADA, M., SATO, M., & YAMAMURA, K. (1973) An experimental study of teratogenic effect of cadmium. Ind. Health, **11**: 127-139.

ITO, K. & SAWAUCHI, K. (1966) Inhibitory effects on cadmium-induced testicular damage by pretreatment with smaller cadmium doses. Okajimas Folia anat. Jpn. **42**: 107-117.

ITOKAWA, Y., ABE, T., & TANAKA, S. (1973) Bone changes in experimental chronic cadmium poisoning. Arch. environ. Health, **26**: 241-244.

ITOKAWA, Y., ABE, T., TABEI, R., & TANAKA, S. (1974) Renal and skeletal lesions in experimental cadmium poisoning. Arch. environ. Health, **28**: 149-154.

ITOKAWA, Y., NISHINO, K., TAKASHIMA, M., NAKATA, T., KAITO, H., OKAMOTO, E., DAIJO, K., & KAWAMURA, J. (1978) Renal and skeletal lesions in experimental cadmium poisoning of rats. Histology and renal function. Environ. Res., **15**: 206-217.

IWAO, S. (1977) Cadmium, lead, copper, and zinc in food, faeces, and organs of humans. Keio J. Med., **26**: 63-78.

IWAO, S., SUGITA, M., & TSUCHIYA, K. (1981a) Some metabolic interrelationships among cadmium, lead, copper, and zinc: results from a field survey in Cd-polluted areas in Japan. I. Dietary intake of the heavy metals. Keio J. Med., **30**: 17-36.

IWAO, S., SUGITA, M., & TSUCHIYA, K. (1981b) Some metabolic interrelationships among cadmium, lead, copper, and zinc: results from a field survey in Cd-polluted areas in Japan. II. Faecal excretion of the heavy metals. Keio J. Med., **30**: 71-82.

JAGNER, D., JOSEFSON, M., & WESTERLUND, S. (1981) Simultaneous determination of cadmium and lead in urine by means of computerized potentiometric stripping analysis. Anal. Chim. Acta, **128**: 155-161.

JAMALL, I.S. & SMITH, J.C. (1986) The effect of dietary selenium on cadmium cardiotoxicity. In: Born, G.V.R., Rensselaer, A.F., Herken, H., & Welch, A.D., ed. Handbook of experimental pharmacology, Berlin, Heidelberg, Springer Verlag, Vol. 80, pp. 351-361.

JAMALL, I.S. & SPROWLS, J.J. (1987) Effects of cadmium and dietary selenium on cytoplasmic and mitochondrial antioxidant defense systems in the heart of rats fed high dietary copper. Toxicol. appl. Pharmacol., **87**: 102-110.

JANSSENS, M. & DAMS, R. (1974) Determination of cadmium in air particulates by flameless atomic absorption spectrometry with a graphite tube. Anal. Chim. Acta, **70**: 25-33.

JAPAN CADMIUM RESEARCH COMMITTEE (1989) [Summary report on studies of health effects of cadmium], Tokyo, Japan Public Health Association, pp. 53-67 (Kankyo Hoken Report No. 56) (in Japanese).

JAPAN PUBLIC HEALTH ASSOCIATION (1968) [A study of the cause of Itai-itai disease], Tokyo, Japan Public Health Association, 73 pp (in Japanese).

JAPAN PUBLIC HEALTH ASSOCIATION (1970) [Study on cadmium intake and accumulation in areas requiring official observation], Tokyo, Japan Public Health Association, 49 pp (in Japanese).

JAPANESE ENVIRONMENT AGENCY (1972) [Results of soil pollution survey in 1971], Tokyo, Japanese Environment Agency, Soil Agricultural Chemical Division, Water Quality Bureau, 129 pp (in Japanese).

JAPANESE ENVIRONMENT AGENCY (1974) [Results of measurement at the national air sampling network stations], Tokyo, Japanese Environment Agency, Air Pollution Control Division, Air Quality Bureau, 159 pp (in Japanese).

JAPANESE ENVIRONMENT AGENCY (1982) [Results of soil pollution survey in 1981], Tokyo, Japanese Government Agency, Soil Agricultural Chemical Division, Water Quality Bureau, pp. 1-29 (in Japanese).

JAPANESE MINISTRY OF AGRICULTURE AND FORESTRY (1973) [Results of a survey for protection measures of soil contamination in 1972.] In: [Materials for soil protection measures], 197 pp (Report No. 44) (in Japanese).

JAPANESE MINISTRY OF HEALTH AND WELFARE (1968) [Opinion of the Ministry with regard to Itai-itai disease and its causes], Tokyo, Japanese Ministry of Health and Welfare (in Japanese).

JAPANESE MINISTRY OF HEALTH AND WELFARE (1971) [Partial revision of health examination methods in the provisional counter-measures against environmental pollution by cadmium], Tokyo, Japanese Ministry of Health and Welfare, Environmental Pollution Section (Unpublished document) (in Japanese).

JAPANESE MINISTRY OF HEALTH AND WELFARE (1983) [The present status of nutrition among the people], Tokyo, Japanese Ministry of Health and Welfare, 56 pp (in Japanese).

JÄRUP, L., ROGENFELT, A., ELINDER, C.-G., NOGAWA, K., & KJELLSTRÖM, T. (1983) Biological half-time of cadmium in the blood of workers after cessation of exposure. Scand. J. Work Environ. Health, **9**: 327-331.

JÄRUP, L., ELINDER, C.G., & SPANG, G. (1988) Cumulative blood-cadmium and tubular proteinuria: a dose-response relationship. Int. Arch. occup. environ. Health, **60**: 223-229.

JAWAID, M., LIND, B., & ELINDER, C.-G (1983) Determination of cadmium in urine by extraction and flameless atomic absorption spectrophotometry. Comparison of urine from smokers and non-smokers of different sex and age. Talanta, **30**: 509-513.

JEFFERY, E.H., NOSEWORTHY, R., & CHERIAN, M.G. (1989) Age dependent changes in metallothionein and accumulation of cadmium in horses. Comp. Biochem. Physiol., **93C**: 327-332.

JIN, T., NORDBERG, G.F., & NORDBERG, M. (1987a) Resistance to acute nephrotoxicity induced by cadmium-metallothionein dependence on pretreatment with cadmium chloride. Pharmacol. Toxicol., **61**: 89-93.

JIN, T., NORDBERG, G.F., & NORDBERG, M. (1987b) Influence of cadmium-metallothionein pretreatment on tolerance of rat kidney cortical cells to cadmium toxicity *in vitro* and *in vivo*. Pharmacol. Toxicol., **60**: 345-349.

JIN, T., NORDBERG, G.F., & NORDBERG, M. (1986) Uptake of cadmium in isolated kidney cells - influence of binding form and *in vivo* pretreatment. J. appl. Toxicol., **6**(6): 397-400.

JOHNSON, M.S. & EATON, J.W. (1980) Environmental contamination through residual trace metal dispersal from derelict lead-zinc mine. J. environ. Qual., **9**(2): 175-179.

JONES, K.C., SYMON, C.J., & JOHNSTON, A.E. (1987) Long-term changes in soil and cereal grain cadmium: studies at Rothamsted Experimental Station. In: Hemphill, D.D., ed. Trace substances in environmental health - XXI. Proceedings of the University of Missouri's 21st Annual Conference on Trace Substances in Environmental Health, St. Louis, Missouri, 25-28 May, 1987, Columbia, Missouri, University of Missouri Press, pp. 450-460.

JORHEM, L., MATTSON, P., & SLORACH, S. (1984) Lead, cadmium, zinc and certain other metals in foods on the Swedish market. Var Föda, **36**: Suppl. 3.

JUST, J. & KELUS, J. (1971) [Cadmium in the air atmosphere of ten selected cities in Poland.] Rocz. Panstw. Zakl. Hig., **22**: 249-256 (in Polish).

KAGI, J.H.R. & KOJIMA, Y. (1987) Metallothionein II. Proceedings of the Second International Meeting on Metallothionein and Other Low Molecular Weight Metal-Binding Proteins, Zurich, 21-24 August, 1985, Basel, Birkhauser Verlag.

KAGI, J.H.R. & NORDBERG, M., ed. (1979) Metallothionein, Basel, Birkhauser Verlag.

KAGI, J.H.R. & VALLEE, B.L. (1960) Metallothionein: a cadmium- and zinc-containing protein from equine renal cortex. I. J. biol. Chem., **235**: 3460-3465.

KAGI, J.H.R. & VALLEE, B.L. (1961) Metallothionein: a cadmium- and zinc-containing protein from equine renal cortex. II. Physiochemical properties. J. biol. Chem., **236**: 2435-2442.

KAHN, H.L. & MANNING, D.C. (1972) Background correction in atomic absorption spectroscopy. Am. Lab., **August**: 51-56.

KAJIKAWA, K., NAKANISHI, I., & KURODA, K. (1981) Morphological changes of the kidney and bone of rats in chronic cadmium poisoning. Exp. mol. Pathol., **34**: 9-24.

KANAI, M., NOMOTO, S., SASAOKA, S., & NAIKI, M. (1971) Clinical significance of urinary excretion of retinol-binding protein in patients with Itai-itai disease. Proc. Symp. Physiol. Pathol., **11**: 194-199.

KANISAWA, M. & SCHROEDER, H.A. (1969) Life term studies on the effect of trace elements on spontaneous tumours in mice and rats. Cancer Res., **21**: 892-895.

KAPLAN, P.D., BLACKSTONE, M., & RICHDALE, N. (1973) Direct determination of cadmium, nickel, and zinc in rat lungs. Arch. environ. Health, **27**: 387-389.

KAR, A.B., DAS, R.P., & KARKUN, J.N. (1959) Ovarian changes in prepubertal rats after treatment with cadmium chloride. Acta biol. med. Germ., **3**: 372-399.

KAR, A.B., DAS, R.P., & MUKERJI, B. (1960) Prevention of cadmium-induced changes in the gonads of rat by zinc and selenium: a study in antagonism between metals in the biological system. Proc. Natl Inst. Sci. India Part B. Biol. Sci., **26**: 40-50.

KARIN, M. & RICHARDS, R.I. (1982) Human metallothionein genes: primary structure of the metallothionein-II gene and a related processed gene. Nature (Lond.), **299**: 797-802.

KARIN, M., SLATER, E.P., & HERSCHMAN, H.R. (1981) Regulation of metallothionein synthesis in HeLa cells by heavy metals and glucocorticoids. J. cell Physiol., **106**: 63-74.

KASUYA, M., TERANISHI, H., KUBOTA, Y., KATO, T., AOSHIMA, K., SAIJO, A., IWATA, K., & KANAI, M. (1986) [Biological meaning and blood and urine low-molecular-weight protein determinations of Itai-itai disease patient and their families, 10 years follow-up study on urine low-molecular weight protein], Tokyo, Japan Public Health Association, pp. 176-180 (Kankyo Hoken Report No. 52) (in Japanese).

KATAGIRI, Y., TATI, M., IWATA, H., & KANAI, M. (1971) [Concentration of cadmium in urine by age.] Med. Biol., **82**: 239-243 (in Japanese).

KATO, T. & ABE, K. (1978) Environmental pollution and health effects - Toyama Prefecture. In: Tsuchiya, K., ed. Cadmium studies in Japan: a review, Amsterdam, Oxford, New York, Elsevier Science Publishers, pp. 225-249.

KAWAI, K. (1978) Experimental studies on animals. Morphological changes of the kidney and effects on bones. In: Tsuchiya, K., ed. Cadmium studies in Japan: a review, Amsterdam, Oxford, New York, Elsevier Science Publishers, pp. 86-97.

KAWAI, M., FUKUDA, K., & KIMURA, M. (1976) Morphological alterations in experimental cadmium exposure with special reference to the onset of renal lesion. In: Nordberg, G.F., ed. Effects and dose-response relationships of toxic metals, Amsterdam, Oxford, New York, Elsevier Science Publishers, pp. 343-370.

KAWAMURA, J., YOSHIDA, O., NISHINO, K., & ITOKAWA, Y. (1978) Disturbances in kidney functions and calcium and phosphate metabolism in cadmium-poisoned rats. Nephron, **20**: 101-110.

KAWANO, S. & KATO, T. (1978) Environmental pollution and health effects - Ishikawa Prefecture. In: Tsuchiya, K., ed. Cadmium studies in Japan: a review, Amsterdam, Oxford, New York, Elsevier Science Publishers, pp. 220-225.

KAWASHIMA, H., NOMIYAMA, H., & NOMIYAMA, K. (1988) Chronic exposure to cadmium did not impair vitamin D metabolism in monkeys. Environ. Res., **46** 48-58.

KAZANTZIS, G. (1956) Respiratory function in men casting cadmium alloys. Part I. Assessment of ventilatory function. Br. J. ind. Med., **13**: 30-36.

KAZANTZIS, G. (1963) Induction of sarcoma in the rat by cadmium sulfide pigment. Nature ( Lond.), **198**: 1213-1214.

KAZANTZIS, G. (1979) Renal tubular dysfunction and abnormalities of calcium metabolism in cadmium workers. Environ. Health Perspect., **28**: 155-160.

KAZANTZIS, G. & ARMSTRONG, B.G. (1982) A mortality study of cadmium workers in the United Kingdom. Scand. J. Work Environ. Health, **8**(1): 157-160.

KAZANTZIS, G. & HANBURY, W.J. (1966) The induction of sarcoma in the rat by cadmium sulfide and by cadmium oxide. Br. J. Cancer, **20**: 190-199.

KAZANTZIS, G., FLYNN, F.V., SPOWAGE, J.S., & TROTT, D.G. (1963) Renal tubular malfunction and pulmonary emphysema in cadmium pigment workers. Q. J. Med., **32**: 165-192.

KAZANTZIS, G., LAM, T.-H., & SULLIVAN, K.R. (1988) Mortality of cadmium-exposed workers. A five-year update. Scand. J. Work Environ. Health, **14**: 220-223.

KAZANTZIS, G., BLANKS, R., & TEATHER, S. (1989) The mortality of cadmium exposed workers. In: Vernet, J.-P., ed. Proceedings of the International Conference on Heavy Metals in the Environment, Geneva, September 1989, Edinburgh, CEP Consultants Ltd, Vol. 1, pp. 304-307.

KELLO, D. & KOSTIAL, K. (1977) Influence of age and milk diet on cadmium absorption from the gut. Toxicol. appl. Pharmacol., **40**: 277-282.

KIDO, T., HONDA, R., TSURITANI, I., YAMAYA, H., ISHIZAKI, M., YAMADA, Y., & NOGAWA, K. (1988) Progress of renal dysfunction in inhabitants environmentally exposed to cadmium. Arch. environ. Health, **43**: 213-217.

KIDO, T., NOGAWA, K., YAMADA, Y., HONDA, R., TSURITANI, I., ISHIZAKI, M., & YAMAYA, H. (1989) Osteopenia in inhabitants with renal dysfunction induced by exposure to environmental cadmium. Int. Arch. occup. environ. Health, **61**: 271-276.

KIMURA, M. & OTAKI, N. (1972) Percutaneous absorption of cadmium in rabbit and hairless mouse. Ind. Health (Tokyo), **10**: 7-10.

KING, E. (1955) An environmental study of casting copper-cadmium alloys. Br. J. ind. Med., **12**: 198-205.

KIPLING, M.D. & WATERHOUSE, J.A.H. (1967) Cadmium and prostatic carcinoma. Lancet, **1**: 730-731.

KIRSCH, H., PADBERG, W., SCHOLZ, A., & ZIMMERMEYER, G. (1982) Cadmium emissions from coal-fired power plants. In: Proceedings of the 3rd International Cadmium Conference, Miami, Florida, 3-5 February, 1981, London, Cadmium Association, pp. 64-68.

KITAMURA, M. (1972) [Absorption and deposition of cadmium, especially in human subjects], Tokyo, Japan Public Health Association, pp. 42-45 (Kankyo Hoken Report No. 11) (in Japanese).

KITAMURA, M. & KOIZUMI, N. (1975) [Results of urine tests for bedridden old persons in Hyogo Prefecture. Report 2], Tokyo, Japan Public Health Association, pp. 135-137 (Kankyo Hoken Report No. 36) (in Japanese).

KITAMURA, S., SUMINO, K., & KAMATANI, N. (1970) [Cadmium concentrations in liver, kidneys, and bones of human bodies.] Jpn. J. public Health, **17**: 177 (in Japanese).

KJELLSTRÖM, T. (1971) A mathematical model for the accumulation of cadmium in human kidney cortex. Nord. Hyg. Tidskr., **53**: 111-119.

KJELLSTRÖM, T. (1977) Accumulation and renal effects of cadmium in men. A dose-response study, Stockholm, Karolinska Institute, 80 pp (Doctoral thesis).

KJELLSTRÖM, T. (1979) Exposure and accumulation of cadmium in populations from Japan, the United States, and Sweden. Environ. Health Perspect., **28**: 169-197.

KJELLSTRÖM, T. (1986) Effects on bone, on vitamin D, and calcium metabolism. In: Friberg, L., Elinder, C.-G., Kjellström, T., & Nordberg, G.F., ed. Cadmium and health: A toxicological and epidemiological appraisal, Boca Raton, Florida, CRC Press, Vol. II, pp. 111-158.

KJELLSTRÖM, T. & NORDBERG, G.F. (1978) A kinetic model of cadmium metabolism in the human being. Environ. Res., **16**: 248-269.

KJELLSTRÖM, T. & PISCATOR, M. (1977) Quantitative analysis of $\beta$-2-microglobulin in urine as an indicator of renal tubular damage by cadmium. Phadedoc No. 1. Diagnostic communications, Uppsala, Sweden, Pharmacia Diagnostics, 21 pp.

KJELLSTRÖM, T., LIND, B., LINNMAN, L., & NORDBERG, G.F. (1974) A comparative study on methods for cadmium analysis of grain with an application to pollution evaluation. Environ. Res., **8**: 92-106.

KJELLSTRÖM, T., LIND, B., LINNMAN, L., & ELINDER, C.-G. (1975a) Variation of cadmium concentration in Swedish wheat and barley. An indicator of changes in daily cadmium intake during the 20th century. Arch. environ. Health, **30**: 321-328.

KJELLSTRÖM, T., TSUCHIYA, K., TOMPKINS, E., TAKABATAKE, E., LIND, B., & LINNMAN, L. (1975b) A comparison of methods for analysis of cadmium in food and biological material. A cooperative study between Sweden, Japan, and the USA. In: Recent advances in the assessment of the health effects of environmental pollution, Luxembourg, Commission of the European Communities, pp. 2197-2213.

KJELLSTRÖM, T., EVRIN, P.E., & RAHNSTER, B. (1977a) Dose-response analysis of cadmium-induced tubular proteinuria. A study of urinary $\beta$-2-microglobulin excretion among workers in a battery factory. Environ. Res., **13**: 303-317.

KJELLSTRÖM, T., SHIROISHI, K., & EVRIN, P.E. (1977b) Urinary $\beta$-2-microglobulin excretion among people exposed to cadmium in the general environment. An epidemiological study in cooperation between Japan and Sweden. Environ. Res., **13**: 318-344.

KJELLSTRÖM, T., BORG, K., & LIND, B. (1978) Cadmium in faeces as an estimator of daily cadmium intake in Sweden. Environ. Res., **15**: 242-251.

KJELLSTRÖM, T., FRIBERG, L., & RAHNSTER, D. (1979) Mortality and cancer morbidity among cadmium-exposed workers. Environ. Health Perspect., **28**: 199-204.

KJELLSTRÖM, T., ELINDER, C.-G., & FRIBERG, L. (1984) Conceptual problems in establishing the critical concentration of cadmium in human kidney cortex. Environ. Res., **33**: 284-295.

KLAASSEN, C.D. & KOTSONIS, F.N. (1977) Biliary excretion of cadmium in the rat, rabbit, and dog. Toxicol. appl. Pharmacol., **41**: 101-112.

KNEIP, T.J., EISENBUD, M., STREHLOW, C.D., & FREUDENTHAL, P.C. (1970) Airborne particulates in New York City. J. Air Pollut. Control Assoc., **20**: 144-149.

KOBAYASHI, J. (1974) Effects of cadmium on calcium metabolism of rats. In: Hemphill, D.D., ed. Trace substances in environmental health - VII, Columbia, Missouri, University of Missouri Press, pp. 295-305.

KODAMA, Y., MORIMOTO, K., & NAKAJIMA, T. (1978) [Studies on the chemical form of heavy metal in oysters.] Eisei Kagaku, **24**: 132-138 (in Japanese).

KOGAN, I.G., GROZDOVA, T.YA., & KHOLIKOVA, T.A. (1978) [Investigation of the mutagenic effect of cadmium chloride on Drosophila melanogaster germ cells.] Genetika, **16**: 2136-2140 (in Russian).

KOGAN, V.H., MALISHEVA, N.B., & FIDAROVA, A.M. (1972) On the role of visceral lesions under chronic cadmium intoxications in the development of bone changes. Trans. Kuban Med. Inst., **36**: 75-78.

KOIRTYOHANN, S.R. & HOPKINS, C.A. (1976) Losses of trace metals during the ashing of biological materials. Analyst, **101**: 870-875.

KOIVISTOINEN, P. (1980) Mineral element composition of Finnish foods: N, K, Ca, Mg, P, S, Fe, Cu, Mn, Zn, Mo, Co, Ni, Cr, F, Se, Si, Rb, Al, B, Br, Hg, As, Cd, Pb, and Ash. Acta agric. Scand., **22**: Suppl.

KOIZUMI, H., YASUDA, K., & KATAYAMA, M. (1977) Atomic absorption spectrophotometry based on the polarization characteristics of the Zeeman effects. Anal. Chem., **49**: 1106-1112.

KOJIMA, S., HAGA, Y., KURIHARA, T., & YAMAWAKI, T. (1975) [Report from Akita Prefecture], Tokyo, Japan Public Health Association, pp. 114-123 (Kankyo Hoken Report No. 36) (in Japanese).

KOJIMA, S., HAGA, Y., KURIHARA, T., & YAMAWAKI, T. (1976) [Report on cadmium pollution in Akita Prefecture], Tokyo, Japan Public Health Association, pp. 114-120 (Kankyo Hoken Report No. 36) (in Japanese).

KOJIMA, S., HAGA, Y., KURIHARA, T., YAMAWAKI, T., & KJELLSTRÖM, T. (1977) A comparison between faecal cadmium and urinary $\beta$-2-microglobulin total protein and cadmium among Japanese farmers. An epidemiological study in cooperation between Japan and Sweden. Environ. Res., **14**: 436-451.

KOLLER, L.D., EXON, J.H., & ROAN, J.G. (1975) Antibody suppression by cadmium. Arch. environ. Health, **30**: 598-601.

KONO, M., YOSHIDA, T., SUGIHARA, H., & HAGINO, N. (1956) [Report on so-called Itai-itai disease - First report.] J. Jap. Soc. Orthopaedis., **30**: 100-101 (in Japanese).

KOPP, S.J., GLONEK, T., ERLANGER, M., PERRY, E.F., PERRY, H.M., Jr, & BARANY, M. (1980a) Cadmium and lead effects on myocardial function and metabolism. J. environ. Pathol. Toxicol., **4**: 205-227.

KOPP, S.J., PERRY, H.M., Jr, GLONEK, T., ERLANGER, M., PERRY, E.F., BARANY, M., & D'AGROSA, L.S. (1980b) Cardiac physiologic-metabolic changes after chronic low-level heavy metal feeding. Am. J. Physiol., **239**: H22-H30.

KOPP, S.J., PERRY, H.M., Jr, PERRY, E.F., & ERLANGER, M. (1983) Cardiac physiologic and tissue metabolic changes following chronic low-level cadmium and cadmium plus lead ingestion in the rats. Toxicol. appl. Pharmacol., **69**: 149-160.

KOROPATNICK, J. & CHERIAN, M.G. (1988) Exposure to different forms of cadmium in mice: differences in metallothionein and alphafetoprotein mRNA induction in liver and kidney. J. biochem. Toxicol., **3**: 159-172.

KORPELA, H., LOUENIVA, R., YRJANHEIKKI, E., & KAUPPILA, A. (1986) Lead and cadmium concentrations in maternal and umbilical cord blood, amniotic fluid, placenta, and amniotic membranes. Am J. Obstet. Gynecol., **155**: 1086-1089.

KOTSONIS, F.N. & KLAASSEN, C.D. (1977) Toxicity and distribution of cadmium administered to rats at sublethal doses. Toxicol. appl. Pharmacol., **41**: 667-680.

KOWAL, N.E. & KRAEMER, D.F. (1982) Urinary cadmium and $\beta$-2-microglobulin levels of persons aged 20-74 years from a sub-sample of the national HANES II survey, 1976-1980. In: Proceedings of the 3rd International Cadmium Conference, Miami, Florida, 3-5 February, 1981, London, Cadmium Association, pp. 119-122.

KOWAL, N.E., JOHNSON, D.E., KRAEMER, D.F., & PAHREN, H.R. (1979) Normal levels of cadmium in diet, urine, blood, and tissues of inhabitants of the United States. J. Toxicol. environ. Health, **5**: 995-1014.

KRAEMER, D.F., LUCAS, J.B., PAHREN, H.R., RYAN, J.A., & KOWAL, N.E. (1979) Cadmium toxicity. Lancet, **1**: 1241-1242.

KRASOVSKII, G.N., VARSHAVSKAYA, S.P., & BORISOV, A.I. (1976) Toxic and gonadotropic effects of cadmium and boron relative to standards for these substances in drinking-water. Environ. Health Perspect., **13**: 69-75.

KRELL, U. & ROECKNER, E. (1988) Model simulation of the atmospheric input of lead and cadmium into the North Sea. Atmos. Environ., **22**(2): 375-381.

KRETZSCHMAR, J.G., DELESPAUL, I., & DE RIJK, T.H. (1980) Heavy levels in Belgium: a five-year survey. Sci. total Environ., **14**: 85-97.

KUHNERT, P.M., KUHNERT, B.R., BOTTOMS, S.F., & ERHARD, P. (1982) Cadmium levels in maternal blood, fetal cord blood, and placental tissues of pregnant women who smoke. Am. J. Obstet. Gynecol., **142**: 1021-1028.

KUHNERT, P.M., KUHNERT, B.R., ERHARD, P., BRASHEAR, W.T., GROH-WARGO, S.L., & WEBSTER, S. (1987a) The effect of smoking on placental and fetal zinc status. Am. J. Obstet. Gynecol., **157**: 1241-1246.

KUHNERT, B.R., KUHNERT, P.M., DEBANNE, S., & WILLIAMS, T.G. (1987b) The relationship between cadmium, zinc and birth weight in pregnant women who smoke. Am. J. Obstet. Gynecol., **157**: 1247-1251.

KUHNERT, B.R., KUHNERT, P.M., & ZARLINGO, T.J. (1988) Associations between placental cadmium and zinc and age and parity in pregnant women who smoke. Obstet. Gynecol., **71**: 67-70.

LAAMANEN, A. (1972) Functions, progress, and prospects for an environmental subarctic base level station. Work environ. Health, **9**: 17-25.

LAFLEUR, P.D., ed. (1976) Accuracy in trace analysis: sampling, sample handling, analysis, Washington, DC, US Department of Commerce, National Bureau of Standards, Vol. 1 and 2, 1304 pp.

LARIONOVA, T.I., ZILLE, L.N., & VOROBJEVA, R.S. (1974) [The influence of barium-cadmium laurate on some fermentative processes in the liver.] Gig. Tr. prof. Zabol., **3**: 50-52 (in Russian).

LAUWERYS, R. & DE WALS, PH. (1981) Environmental pollution by cadmium and mortality from renal diseases. Lancet, **1**: 383.

LAUWERYS, R., BUCHET, J.P., ROELS, H., BROUWERS, J., & STANESCU, D. (1974a) Epidemiological survey of workers exposed to cadmium. Arch. environ. Health, **28**: 145-148.

LAUWERYS, R., BUCHET, J.P., & ROELS, H. (1974b) Effets subcliniques de l'exposition humaine au cadmium. In: Proceedings of the International Symposium on Problems of the Contamination of Man and His Environment by Mercury and Cadmium, Luxembourg, Commission of the European Communities, pp. 447-461.

LAUWERYS, R., BUCHET, J.P., & ROELS, H. (1974c) Dosage du cadmium et du plomb dans le sang et les urines par spectrophotométrie d'absorption atomique précedée d'une chromatographie sur résine échangeuse d'ions. In: Proceedings of the International Symposium on Problems of the Contamination of Man and His Environment by Mercury and Cadmium, Luxembourg, Commission of the European Communities, pp. 231-238.

LAUWERYS, R., BUCHET, J.P., ROELS, H., BERLIN, A., & SMEETS, J. (1975) Intercomparison program of lead, mercury, and cadmium analysis in blood, urine, and aqueous solutions. Clin. Chem., **21**: 551-557.

LAUWERYS, R., BUCHET, J.P., & ROELS, H. (1976) The relationship between cadmium exposure or body burden and the concentration of cadmium in blood and urine in man. Int. Arch. occup. environ. Health, **36**: 275-285.

LAUWERYS, R., BUCHET, J.P., ROELS, H., & HUBERMONT, G. (1978) Placental transfer of lead, mercury, cadmium and carbon monoxide in women. I. Comparison of the frequency distributions of the biological indices in maternal and umbilical cord blood. Environ. Res., **15**: 278-290.

LAUWERYS, R., ROELS, H., BUCHET, J.P., BERNARD, A., & STANESCU, D. (1979a) Investigations on the lung and kidney functions in workers exposed to cadmium. Environ. Health Perspect., **28**: 137-145.

LAUWERYS, R., ROELS, H., REGNIERS, M., BUCHET, J.P., BERNARD, A., & GORET, A. (1979b) Significance of cadmium concentration in blood and in urine in workers exposed to cadmium. Environ. Res., **20**: 375-391.

LAUWERYS, R., ROELS, H., BERNARD, A., & BUCHET, J.P. (1980a) Renal response to cadmium in a population living in a non-ferrous smelter area in Belgium. Int. Arch. occup. environ. Health, **45**: 271.

LAUWERYS, R., BUCHET, J.P., ROELS, H., BERNARD, A., CHETTLE, D.R., HARVEY, T.C., & ALHADDAD, I.K. (1980b) Biological significance of cadmium concentration in blood and urine and their application in monitoring workers exposed to cadmium. In: Proceedings of the 2nd International Cadmium Conference, Cannes, 6-8 February, 1977, London, Cadmium Association, pp. 164-167.

LAUWERYS, R., BUCHET, J.P., ROELS, H., VIAU, C., BERNARD, A., BRUAUX, P., CLAEYS-THOREAU, F., & RONDIA, D. (1984a) Potential risk of cadmium for the general population: Update of Liege study. In: Proceedings of the 4th International Cadmium Conference, Munich, 2-4 March, 1983, London, Cadmium Association, pp. 113-114.

LEHMAN, L.D. & POISNER, A.M. (1984) Induction of metallothionein synthesis in cultured human trophoblasts by cadmium and zinc. J. Toxicol. environ. Health, **14**: 419-432.

LEHNERT, G., VON, SCHALLER, K.H., & HAAS, T. (1968) [Cadmium analysis of serum and urine using atomic absorption spectrometry.] Z. klin. Chem. klin. Biochem., **6**: 174-176 (in German).

LEMEN, R.A., LESS, J.S., WAGONER, J.K., & BLEJER, H.P. (1976) Cancer mortality among cadmium production workers. Ann. NY Acad Sci., **271**: 273-279.

LENER, J. & MUSIL, J. (1971) Cadmium influence on the excretion of sodium by kidneys. Experientia (Basel), **27**: 902.

L'EPEE, P., LAZARINI, H., FRANCHOME, J., N'DOKY, T., & LARRIVET, C. (1968) Contribution à l'étude de l'intoxication cadmique. Arch. Mal. prof. Méd. Trav. Sécur. soc., **29**: 485-490.

LEPOW, M.L., BRUCKMAN, L., RUBINO, R.A., MARKOWITZ, S., GILLETTE, M., & KAPISH, J. (1974) Role of airborne lead in increased body burden of lead in Hartford children. Environ. Health Perspect., **7**: 99-102.

LEVY, L.S. & CLACK, J. (1975) Further studies on the effect of cadmium on the prostate gland. I. Absence of prostatic changes in rats given oral cadmium sulfate for two years. Ann. occup. Hyg., **17**: 205-211.

LEVY, L.S., ROE, F.J.C., MALCOLM, D., KAZANTIZIS, G., CLACK, J., & PLATT, H.S. (1973) Absence of prostatic changes in rats exposed to cadmium. Ann. occup. Hyg., **16**: 111-118.

LEVY, L.S., CLACK, J., & ROE, F.J.C. (1975) Further studies on the effect of cadmium on the prostate gland. II. Absence of prostatic changes in mice given oral cadmium sulfate for eighteen months. Ann. occup. Hyg., **17**: 213-220.

LEWIS, G.P., JUSKO, W.J., COUGHLIN, L.L., & HARTZ, S. (1972) Cadmium accumulation in man: influence of smoking, occupation, alcoholic habit, and disease. J. chron. Dis., **25**: 717-726.

LIKUTOVA, I.V. (1989) [Specific effect of inorganic cadmium compounds on health and its hygienic assessment.] Abstract from a dissertation for degree of Candidate of Medical Sciences, Moscow, Central Institute for the Advanced Training of Physicians, 21 pp (in Russian).

LIKUTOVA, I.V. & BELOVA, E.A. (1987) [Specific action of cadmium in its peroral intake with water.] Gig. i Sanit., **1987**(6): 70-72 (in Russian).

LINDBERG, S.E., TURNER, R.R., & LOVETT, G.M. (1982) Processes of atmospheric deposition of metals and acids to forests. Conference Proceeding 75. Annual Meeting on the Air Pollution Control Association, New Orleans, 20 June, 1982, Pittsburgh, Pennsylvania, Air Pollution Control Association, 14 pp.

LINNMAN, L., ANDERSSON, A., NILSSON, K.O., LIND, B., KJELLSTRÖM, T., & FRIBERG, L. (1973) Cadmium uptake by wheat from sewage sludge used as a plant nutrient source. Arch. environ. Health, **27**: 45-47.

LORENTZON, R. & LARSSON, S.E. (1977) Vitamin D metabolism in adult rats at low and normal calcium intake and the effect of cadmium exposure. Clin. Sci. mol. Med., **53**: 439-446.

LOSER, E. (1980) A 2-year oral carcinogenicity study with cadmium on rats. Cancer Lett., **9**: 191-198.

LUCAS, P.A., JARIVALLA, A.G., JONES, J.H., GOUGH, J., & VALE, P.T. (1980) Total cadmium fume inhalation. Lancet, **2**(8187): 205.

LUCIS, O.J., LUCIS, R., & ATERMAN, K. (1972) Tumorigenesis by cadmium. Oncology, **26**: 53-76.

LUFKIN, N.H. & HODGES, F.T. (1944) Cadmium poisoning. Report of outbreak. US nav. med. Bull., **43**: 1273-1276.

LUND, L.J., BETTY, E.E., PAGE, A.L., & ELLIOTT, R.A. (1981) Occurrence of naturally high cadmium levels in soils and its accumulation by vegetation. J. environ. Qual., **10**: 551-556.

LUNDGREN, G. (1976) Direct determination of cadmium in blood with a temperature-controlled heated graphite-tube atomizer. Talenta, **23**: 309-312.

MAACK, T., JOHNSON, V., KAN, S.T., FIGUEIREDO, J., & SIGULEM, D. (1979) Renal filtration, transport, and metabolism of low molecular weight proteins. A review. Kidney Int., **16**: 251-270.

MACFARLAND, H.N. (1979) Pulmonary effects of cadmium. In: Mennear, J., ed. Cadmium toxicity, New York, Basel, Marcel Dekker, Inc., pp. 113-132.

MCINNES, G. (1979) Multi-element survey: analysis of the first two years' results. Warren Spring Laboratory, United Kingdom Department of Industry, 44 pp (Report No. LR305 AP).

MCKENZIE, J., KJELLSTRÖM, T., & SHARMA, R.P. (1982) Cadmium intake metabolism and effects in people with a high intake of oysters in New Zealand, Washington, DC, US Environmental Protection Agency, 210 pp.

MCLELLAN, J.S., THOMAS, B.J., & FREMLIN, J.H. (1975) Cadmium: its *in vivo* detection in man. Phys. Med. Biol., **20**: 88-95.

MCLELLAN, J.S., FLANAGAN, P.R., CHAMBERLAIN, M.J., & VALBERG, L.S. (1978) Measurements of dietary cadmium absorption in humans. J. Toxicol. environ. Health, **4**: 131-138.

MAHAFFEY, K.R., CORNELIUSSEN, P.E., JELINEK, C.F., & FIORINO, J.A. (1975) Heavy metal exposure from foods. Environ. Health Perspect., **12**: 63-69.

MANGELSON, N.F., HILL, M.W., NIELSON, K.K., EATOUGH, D.J., CHRISTENSEN, J.J., IZATT, R.M., & RICHARDS, D.O. (1979) Proton-induced X-ray emission analysis of Pima Indian autopsy tissues. Anal. Chem., **51**: 1187-1194.

MARC (1986) Biological monitoring of environmental contaminants (plants), London, Monitoring and Assessment Research Centre, Chelsea College, University of London, 247 pp (MARC Report Number 32).

MARGOSHES, M. & VALLEE, B.L. (1957) A cadmium kidney protein from equine kidney cortex. J. Am. Chem. Soc., 79: 4813-4814.

MARTIN, J.H. & BROENKOW, W.W. (1975) Cadmium in plankton: elevated concentrations off Baja California. Science, 190: 884-885.

MARTIN, J.H., ELLIOTT, P.D., ANDERLINI, V.C., GIRVIN, D., JACOBS, S.A., RISEBROUGH, R.W., DELONG, R.L., & GILMARTIN, W.G. (1976) Mercury-selenium-bromine imbalance in premature parturient California sea lions. Mar. Biol., 35: 91-104.

MASON, H.J., DAVISON, A.G., & WRIGHT, A.L. (1988) Relations between liver cadmium, cumulative exposure, and renal function in cadmium alloy workers. Br. J. ind. Med., 45: 793-802.

MASON, K.E. & YOUNG, J.O. (1967) Effectiveness of selenium and zinc in protecting against cadmium-induced injury of rat testis. In: Muth, O.H., ed. Symposium: Selenium in Biomedicine, Westport, Connecticut, Avi, pp. 383-394.

MATERNE, D., LAUWERYS, R., BUCHET, J.P., ROELS, H., BROUWERS, J., & STANESCU, D. (1975) Investigations sur les risques résultant de l'exposition au cadmium dans deux entreprises de production et deux entreprises d'utilisation du cadmium. Cah. Méd. Trav., 12: 1-76.

MATSUSAKA, N., TANAKA, M., NISHIMURA, Y., YUYAMA, A., & KOBAYASHI, H. (1972) [Whole body retention and intestinal absorption of $^{115}$Cd in young and adult mice.] Med. Biol., 85: 275-279 (in Japanese).

MERALI, Z., KACEW, S., & SINGHAL, R.L. (1974) Response of hepatic carbohydrate and cyclic AMP metabolism to cadmium treatment in rats. Can. J. Physiol. Pharmacol., 53: 174-184.

MERANGER, J.C., SUBRAMANIAN, K.S., & CHALIFOUX, C. (1981) Metals and other elements. Survey for cadmium, cobalt, chromium, copper, nickel, lead, zinc, calcium, and magnesium in Canadian drinking-water supplies. J. Assoc. Off. Anal. Chem., 64: 44-53.

MILLER, R.K. (1986) Placental transfer and function. The interface for drugs and chemicals in the conceptus. In: Fabro, S. & Scialli, A.S., ed. Drug and chemical action in pregnancy, New York, Basel, Marcel Dekker, Inc., pp. 123-152.

MILLS, C.F. & DELGARNO, A.C. (1972) Copper and zinc status of ewes and lambs receiving increased dietary concentration of cadmium. Nature ( Lond.), 239: 171-173.

MOGENSEN, C.E. & SOLLING, K. (1977) Studies on renal tubular protein reabsorption: partial and near complete inhibition by certain amino acids. Scand. J. clin. lab. Invest., 37: 477-486.

MOHD, M.A., MURTHY, R.C., & CHANDRA, S.V. (1986) Developmental and long-term neurobehavioral toxicity of low-level *in utero* Cd exposure in rats. Toxicol. Teratol., **8**: 463.

MORGAN, J.M. (1970) Cadmium and zinc abnormalities in bronchogenic carcinoma. Cancer, **25**: 1394-1398.

MORGAN, J.M., BURCH, H.B., & WATKISS, J.B. (1971) Tissue cadmium and zinc content in emphysema and bronchogenic carcinoma. J. chron. Dis., **24**: 107-110.

MORITSUGI, M. & KOBAYASHI, J. (1964) Study on trace elements in biomaterials. II. Cadmium content in polished rice. Ber. Ohara Inst. Landwirtsch. Biol. Okayama Univ., **12**: 145-158.

MOSCHANDREAS, D.J. (1981) Exposure to pollutants and daily time budgets of people. Bull. NY Acad. Med., **57**: 845-859.

MULFORD, C.E. (1966) Solvent extraction techniques for atomic absorption spectoscopy. At. Absorpt. Newsl., **5**: 88-90.

MULVIHILL, J.E., GAMM, S.H., & FERM, V.H. (1970) Facial formation in normal and cadmium-treated golden hamsters. J. Embryol. exp. Morphol., **24**: 393-403.

MURASE, H., INOUE, T., SAGA, M., IWATA, T., & KUBOTA, K. (1974) [Microscopic pathology of dogs chronically exposed to oral cadmium], Tokyo, Japan Public Health Association, pp. 72-73 (Kankyo Hoken Report No. 31) (in Japanese).

MUSKETT, C.J., ROBERTS, L.H., & PAGE, B.J. (1979) Cadmium and lead pollution from secondary metal refinery operations. Sci. total Environ., **11**: 73-87.

NACKOWSKI, S.B., PUTNAM, R.D., ROBBINS, D.A., VARNER, M.O., WHIT, L.D., & NELSON, K.W. (1977) Trace metal contamination of evacuated blood collection tubes. Am. Ind. Hyg. Assoc. J., **38**: 503-508.

NAKAGAWA, S. (1960) [A study of osteomalacia in Toyama Prefecture (so-called Itai-itai disease).] J. Radiol. Phys. Ther. Univ. Kanazawa, **56**: 1-51 (in Japanese with English summary).

NAKANO, M., AOSHIMA, K., KATOH, T., TERANISHI, H., & KATOH, T. (1987) Severity of tubular brush border damage in a cadmium-polluted area (Jinzu River basin): clinical role of urinary trehalase. Environ. Res., **44**: 161-168.

NAZARI, G., FAVINO, A., & POZZI, U. (1967) [Effects of a single subcutaneous injection of cadmium chloride in male rats.] Riv. Anat. Pathol. Oncol., **31**: 251-271 (in Italian).

NEEB, R. & WAHDAT, F. (1974) [About inverse voltammetry analysis of cadmium in aerosols.] Z. anal. Chem., **269**: 275-279 (in German).

NICAUD, P., LAFITTE, A., & GROS, A. (1942) Les troubles de l'intoxication chronique par le cadmium. Arch. Mal. prof., **4**: 192-202.

NICHOLSON, J.K. & OSBORN, O. (1983) Kidney lesions in pelagic seabirds with high tissue levels of cadmium and mercury. J. Zool. (Lond.), **200**: 88-118.

NIELSEN, S.A. (1975) Cadmium in New Zealand dredge oysters, geographic distribution. Int. J. environ. anal. Chem., **4**: 1-7.

NISE, G. & VESTERBERG, O. (1978) [Sampling vessels influence on content of heavy metals (Cd, Cr, Hg, Ni, and Pb) in blood and urine samples.] Läkartidningen, **75**: 19-20 (in Swedish).

NISHIYAMA, K. & NORDBERG, G. (1972) Adsorption and elution of cadmium on hair. Arch. environ. Health, **25**: 92-96.

NISHIYAMA, S., NAKAMURA, K., & KONISHI, Y. (1986) Blood pressure and urinary sodium and potassium excretion in cadmium-treated male rats. Environ. Res., **40**: 357-364.

NOGAWA, K. (1981) Itai-itai disease and follow-up studies. In: Nriagu, J.O., ed. Cadmium in the environment. II. Health effects, New York, John Wiley and Sons, pp. 1-38.

NOGAWA, K. & ISHIZAKI, A. (1979) A comparison between cadmium in rice and renal effects among inhabitants of the Jinzu River Basin. Environ. Res., **18**: 410-420.

NOGAWA, K., ISHIZAKI, A., FUKUSHIMA, M., SHIBATA, I., & HAGINO, N. (1975) Studies on the women with acquired Fanconi syndrome observed in the Ichi River basin polluted by cadmium. Environ. Res., **10**: 280-307.

NOGAWA, K., ISHIZAKI, A., & KAWANO, S. (1978) Statistical observations of the dose-response relationships of cadmium based on epidemiological studies in the Kakehashi River basin. Environ. Res., **15**: 185-198.

NOGAWA, K., KOBAYASHI, E., HONDA, R., & ISHIZAKI, A. (1979) Clinico-chemical studies on chronic cadmium poisoning. I. Results of urinary examinations. Jpn. J. Hyg., **34**: 407-414.

NOGAWA, K., KOBAYASHI, E., HONDA, R., ISHIZAKI, A., KAWANO, S., & MATUSDA, H. (1980) Renal dysfunction of inhabitants in a cadmium-polluted area. Environ. Res., **23**: 13-23.

NOGAWA, K., KOBAYASHI, E., & KONISHI, F. (1981a) Comparison of bone lesions in chronic cadmium intoxication and vitamin D deficiency. Environ. Res., **24**: 233-249.

NOGAWA, K., KAWANO, S., & NISHI, M. (1981b) Mortality study of inhabitants in a cadmium-polluted area with special reference to low molecular weight proteinuria. In: Heavy metals in the environment, Edinburgh, CEP Consultants Ltd, pp. 538-540.

NOGAWA, K., KIDO, T., YAMADA, Y., TSURITANI, I., HONDA, R., ISHIZAKI, N., & TERAHATA, K. (1984) Alpha-1-microglobulin in urine as an indicator of renal tubular damage caused by environmental cadmium exposure. Toxicol. Lett., **22**: 63-68.

NOGAWA, K., TSURITANI, I., YAMADA, Y., KIDO, T., HONDA, R., ISHIZAKI, M., & KURIHARA, T. (1986) Sister-chromatid exchanges in the lymphocytes of people exposed to environmental cadmium. Toxicol. Lett., **32**: 283-288.

NOGAWA, K., TSURITANI, I., KIDO, T., HONDA, R., YAMADA, Y., & ISHIZAKI, M. (1987) Mechanism for bone disease found in inhabitants environmentally exposed to cadmium: decreased serum 1α,25-dihydroxyvitamin D level. Int. Arch. occup. environ. Health, **59**: 21-30.

NOGAWA, K., HONDA, R., KIDO, T., TSURITANI, I., YAMADA, Y., ISHIZAKI, M., & YAMAYA, H. (1989) A dose-response analysis of cadmium in the general environment with special reference to total cadmium intake limit. Environ. Res., **48**: 7-16.

NOMIYAMA, K. (1973) Early signs of cadmium intoxication in rabbits. Toxicol. appl. Pharmacol., **24**: P625-P635.

NOMIYAMA, K. (1974) [Studies concerning cadmium metabolism and the etiology of poisoning], Tokyo, Japan Public Health Association, pp. 53-59 (Kankyo Hoken Report No. 31) (in Japanese).

NOMIYAMA, K. (1977) Does a critical concentration of cadmium in human renal cortex exist? J. Toxicol. environ. Health, **3**: 607-609.

NOMIYAMA, K. (1978) Experimental studies on animals: *in vivo* experiments. In: Tsuchiya, K., ed. Cadmium studies in Japan: a review, Amsterdam, Oxford, New York, Elsevier Science Publishers, pp. 47-86.

NOMIYAMA, K. (1986) The chronic toxicity of cadmium: influence of environmental and other variables. In: Born, G.V.R., Rensselaer, A.F., Herken, H., & Welch, A.D., ed. Handbook of experimental pharmacology, Berlin, Heidelberg, Springer-Verlag, Vol. 80, pp. 101-133.

NOMIYAMA, K. & FOULKES, E.C. (1977) Reabsorption of filtered cadmium metallothionein in the rabbit kidney. Proc. Soc. Exp. Biol. Med., **156**: 97-99.

NOMIYAMA, K. & NOMIYAMA, H. (1976a) Mechanisms of urinary excretion of cadmium: experimental studies in rabbits. In: Nordberg, G.F., ed. Effects and dose-response relationships of toxic metals, Amsterdam, Oxford, New York, Elsevier Science Publishers, pp. 371-379.

NOMIYAMA, K. & NOMIYAMA, H. (1976b) [Long-term experiment with 300 ppm oral cadmium exposure to rabbits. Intermediate report after 12 months], Tokyo, Japan Public Health Association, pp. 161-173 (Kankyo Hoken Report No. 38) (in Japanese).

NOMIYAMA, K. & NOMIYAMA, H. (1982) Tissue metallothionein in rabbits chronically exposed to cadmium, with special reference to the critical concentration of cadmium in the renal cortex. In: Foulkes, E.C., ed. Biological roles of metallothionein, Amsterdam, Oxford, New York, Elsevier Science Publishers, pp. 47-67.

NOMIYAMA, K. & NOMIYAMA, H. (1988) Health effect of six years of dietary cadmium (cadmium contaminated rice) in monkeys. In: Essential and toxic trace elements in human health and disease, New York, Alan K. Liss, Inc., pp. 589-609.

NOMIYAMA, K., SUGATA, Y., YAMAMOTO, A., & NOMIYAMA, H. (1975) Effects of dietary cadmium on rabbits. I. Early signs of cadmium intoxication. Toxicol. appl. Pharmacol., **31**: 4-12.

NOMIYAMA, K., SUGATA, Y., NOMIYAMA, H., & YAMAMOTO, A. (1976) Dose-response relationship for cadmium. In: Nordberg, G.F., ed. Effects and dose-response relationships of toxic metals, Amsterdam, Oxford, New York, Elsevier Science Publishers, pp. 380-385.

NOMIYAMA, K., NOMIYAMA, H., YOTORIYAMA, M., & TAGUCHI, T. (1978a) Some recent studies on the renal effects of cadmium. In: Proceedings of the First International Cadmium Conference, San Francisco, California, 31 January-2 February, 1977, London, Cadmium Association, pp. 186-194.

NOMIYAMA, K., NOMIYAMA, H., & TAGUCHI, T. (1978b) Effects of environmental temperatures on the toxicity of cadmium in mice. In: Proceedings of the 8th Asian Conference on Occupational Health, Tokyo, Japan Industrial Safety Association, pp. 221-225.

NOMIYAMA, K., NOMIYAMA, H., NOMURA, M., TAGUCHI, T., MATSUI, J., YOTORIYAMA, M., AKAHORI, F., IWAO, S., KOIZUMI, N., MASAOKA, T., KITAMURA, S., TSUCHIYA, K., SUZUKI, T., & KOBAYASHI, K. (1979) Effects of dietary cadmium on rhesus monkeys. Environ. Health Perspect., **28**: 223-243.

NOMIYAMA, K., NOMIYAMA, H., AKAHORI, F., & MASAOKA, T. (1982a) Cadmium health effects in monkeys with special reference to the critical concentration of cadmium in the renal cortex. In: Proceedings of the 3rd International Cadmium Conference, Miami, Florida, 3-5 February, 1981, London, Cadmium Association, pp. 151-156.

NOMIYAMA, K., NOMIYAMA, H., YOTORIYAMA, M., & MATSUI, K. (1982b) Sodium dodecyl sulfate acrylamide gel electrophoretic studies on low molecular weight proteinuria, an early sign of cadmium health effects in rabbits. Ind. Health, **20**: 11-18.

NOMIYAMA, K., NOMIYAMA, H., YOTORIYAMA, M., AKAHORI, F., & MASAOKA, T. (1984) Some comments and proposals on dose and effects for estimating critical concentration of cadmium in the renal cortex. In: Proceedings of the Fourth International Cadmium Conference, London, Cadmium Association, pp. 167-171.

NOMIYAMA, K., AKAHORI, F., MASAOKA, T., ARAI, S., NOMURA, Y., YOTORIYAMA, M., KOBAYASHI, K., SUZUKI, T., KAWASHIMA, H., & ONOZAWA, A. (1987) [Dose effect and dose response relationships of 9 years dietary cadmium in rhesus monkeys], Tokyo, Japan Public Health Association, pp. 1-86 (Kankyo Hoken Report No. 53) (in Japanese).

NOMURA, T., KIMURA, M., TANIOKA, Y., TANIMOTO, Y., SAEGUSA, J., KOTAKI, N., TAZAKI, H., HATA, R., DEGUCHI, N., SUDA, T., & YOSHIKI, S. (1988) [Influence of nutritional factors on cadmium-administered monkeys], Tokyo, Japan Public Health Association, pp. 42-122 (Kankyo Hoken Report No. 54) (in Japanese).

NORDBERG, G.F. (1971) Effects of acute and chronic cadmium exposure on the testicles of mice. With special reference to protective effects of metallothionein. Environ. Physiol. Biochem., **1**: 171-187.

NORDBERG, G.F. (1972) Cadmium metabolism and toxicity. Environ. Physiol. Biochem., **2**: 7-36.

NORDBERG, G.F. (1975) Effects of long-term cadmium exposure on the seminal vesicles of mice. J. Reprod. Fertil., **45**: 165-167.

NORDBERG, G.F. & NISHIYAMA, K. (1972) Whole-body and hair retention of cadmium in mice. Arch. environ. Health, **24**: 209-214.

NORDBERG, G.F. & NORDBERG, M. (1988) Biological monitoring of toxic metals, New York, London, Plenum Press, pp. 151-168.

NORDBERG, G.F. & PISCATOR, M. (1972) Influence of long-term cadmium exposure on urinary excretion of protein and cadmium in mice. Environ. Physiol. Biochem., **2**: 37-49.

NORDBERG, G.F., PISCATOR, M., & LIND, B. (1971) Distribution of cadmium among protein fractions of mouse liver. Acta pharmacol. toxicol., **29**: 456-470.

NORDBERG, G.F., SLORACH, S., & STENSTROM, T. (1973) [Cadmium poisoning caused by a cooled soft drink machine.] Läkartidningen, **70**: 601-604 (in Swedish with English summary).

NORDBERG, G.F., GOYER, R.A., & NORDBERG, M. (1975) Comparative toxicity of cadmium metallothionein and cadmium chloride on mouse kidney. Arch. Pathol., **99**: 192-197.

NORDBERG, G.F., ROBERT, K.-H., & PANNONE, M. (1977) Pancreatic and biliary excretion of cadmium in the rat. Acta pharmacol. toxicol., **41**: 84-88.

NORDBERG, G.F., GARVEY, J.S., & CHANG, C.C. (1982) Metallothionein in plasma and urine of cadmium workers. Environ. Res., **28**: 179-182.

NORDBERG, M. (1978) Studies on metallothionein and cadmium. Environ. Res., **15**: 381-404.

NORDBERG, M. (1984) General aspects of cadmium transport, uptake, and metabolism by the kidney. Environ. Health Perspect., **54**: 13-20.

NORDBERG, M. & NORDBERG, G.F. (1975) Distribution of metallothionein-bound cadmium and cadmium chloride in mice: preliminary studies. Environ. Health Perspect., **12**: 103-108.

NORDBERG, M., NUOTTANIEMI, I., CHERIAN, M.G., NORDBERG, G.F., KJELLSTRÖM, T., & GARVEY, J.S. (1986) Characterization studies on the cadmium-binding proteins from two species of New Zealand oysters. Environ. Health Perspect., **65**: 57-62.

NRIAGU, J.O. (1979) Global inventory of natural and anthropogenic emissions of trace metals to the atmosphere. Nature (Lond.), **279**: 409.

NRIAGU, J.O. & PACYNA, J.M. (1988) Quantitative assessment of worldwide contamination of air, water and soils by trace metals. Nature (Lond.), **333**: 134-139.

NWANKWO, J.N., ELINDER, C.G., PISCATOR, M., & LIND, B. (1977) Cadmium in Zambian cigarettes: an interlaboratory comparison in analysis. Zambia J. Sci. Technol., **2**(4): 1-4.

OBERDORSTER, G. (1988) Deposition and retention modeling of inhaled cadmium in rat and human lung: An example for extrapolation of effects and risk estimation. In: Crapo, J.D., Miller, F.J., Smolko, E.D., Graham, J.A., & Hayes, A.W., ed. Extrapolation of dosimetric relationships for inhaled particles and gases, New York, London, Academic Press, pp. 345-370.

OHANIAN, E.V. & IWAI, J. (1980) Etiological role of cadmium in hypertension in an animal model. J. environ. Pathol. Toxicol., **3**: 229-241.

OHMOMO, Y. & SUMIYA, M. (1981) Estimation of heavy metal intake from agricultural products. In: Kitagishi, K. & Yamane, J., ed. Heavy metal pollution in soils of Japan, Tokyo, Japan Scientific Societies Press, pp. 235-244.

OLDIGES, H., HOCHRAINER, D., & GLASER, U. (1989) Long-term inhalation study with Wistar rats and four cadmium compounds. Toxicol. environ. Chem., **19**: 217-222.

O'RIORDAN, M.L., HUGHES, E.G., & EVANS, H.J. (1978) Chromosome studies on blood lymphocytes of men occupationally exposed to cadmium. Mutat. Res., **58**: 305-311.

PAGE, A.L. (1981) Cadmium in soils and its accumulation by food crops. In: Proceedings of an International Conference on Heavy Metals in the Environment, Amsterdam, September 1981, Edinburgh, CEP Consultants, Ltd, pp. 206-213.

PAGE, A.L., BINGHAM, F.T., & SHANG, A.C. (1981) Cadmium. In: Lepp, N.W., ed. Effect of heavy metal pollution on plants, Barking, Essex, Applied Science Publishers, Vol. 1, pp. 77-109.

PANEMANGALORE, M., BANERJEE, D., ONOSAKA, S., & CHERIAN, M.G. (1983) Changes in the intracellular accumulation and distribution of metallothionein in rat liver and kidney during postnatal development. Dev. Biol., **97**: 95-102.

PARIZEK, J. (1957) The destructive effect of cadmium ion on testicular tissue and its prevention by zinc. J. Endocrinol., **15**: 56-63.

PARIZEK, J. (1960) Sterilization of the male by cadmium salts. J. Reprod. Fertil., **1**: 294-309.

PARIZEK, J. (1964) Vascular changes at sites of oestrogen biosynthesis produced by parenteral injection of cadmium salts: the destruction of placenta by cadmium salts. J. Reprod. Fertil., **7**: 263-265.

PARIZEK, J. & ZAHOR, Z. (1956) Effects of cadmium salts on testicular tissue. Nature (Lond.), **177**: 1036-1037.

PARIZEK, J., OSTADALOVA, I., BENES, I., & PITHA, J. (1968a) The effect of a subcutaneous injection of cadmium salts on the ovaries of adult rats in persistent oestrus. J. Reprod. Fertil., **17**: 559-562.

PARIZEK, J., OSTADALOVA, I., BENES, I., & BABICKY, A. (1968b) Pregnancy and trace elements: the protective effect of compounds of an essential trace element, selenium, against the peculiar toxic effects of cadmium during pregnancy. J. Reprod. Fertil., **16**: 507-509.

PARZYCK, D.C., SHAW, S.M., KESSLER, W.V., VETTER, R.J., VAN SICKLE, D.C., & MAYES, R.A. (1978) Fetal effects of cadmium in pregnant rats on normal and zinc-deficient diets. Bull. environ. Contam. Toxicol., **19**: 206-214.

PATON, G.R. & ALLISON, A.C. (1972) Chromosome damage in human cell cultures induced by metal salts. Mutat. Res., **16**: 332-336.

PERRY, H.M., Jr & ERLANGER, M.W. (1973) Elevated circulating renin activity in rats following doses of cadmium known to induce hypertension. J. lab. clin. Med., **82**: 399-405.

PERRY, H.M., Jr & ERLANGER, M.W. (1974) Metal-induced hypertension following chronic feeding of low doses of cadmium and mercury. J. lab. clin. Med., **83**: 541-547.

PERRY, H.M., Jr & ERLANGER, M.W. (1982) Effect of diet on increases in systolic pressure induced in rats by chronic cadmium feeding. J. Nutr., **112**: 1983-1989.

PERRY, H.M., Jr & KOPP, S.J. (1983) Does cadmium contribute to human hypertension? Sci. total Environ., **26**: 223-232.

PERRY, H.M., Jr, ERLANGER, M.W., YUNICE, A., SCHOEPFLE, E., & PERRY, E.F. (1970) Hypertension and tissue metal levels following intravenous cadmium, mercury, and zinc. Am. J. Physiol., **219**: 755-761.

PERRY, H.M., Jr, PERRY, E.F., & PURIFOY, J.E. (1971) Antinatriuretic effect of intramuscular cadmium in rats. Proc. Soc. Exp. Biol. Med., **136**: 1240-1244.

PERRY, H.M., Jr, ERLANGER, M.W., & PERRY, E.F. (1976) Limiting conditions for the induction of hypertension in rats by cadmium. In: Hemphill, D.D., ed. Trace substances in environmental health - VIII, Columbia, Missouri, University of Missouri Press, pp. 51-57.

PERRY, H.M., Jr, ERLANGER, M., & PERRY, E.F. (1977) Hypertension following chronic, very low-dose cadmium feeding. Proc. Soc. Exp. Biol. Med., **156**: 173-176.

PETERING, D.H., LOFTSGAARDEN, J., SCHNEIDER, J., & FOWLER, B. (1984) Metabolism of cadmium, zinc and copper in the rat kidney: the role of metallothionein and other binding sites. Environ. Health Perspect., **54**: 73-81.

PETERING, H.G., JOHNSON, M.A., & STEMMER, K. (1971) Studies of zinc metabolism in the rat. I. Dose-response effects of cadmium. Arch. environ. Health, **23**: 93-96.

PETERING, H.G., CHOUDHURY, H., & STEMMER, K.L. (1979) Some effects of oral ingestion of cadmium on zinc, copper, and iron metabolism. Environ. Health Perspect., **28**: 97-106.

PETERSON, P.A. (1971) Characteristics of a vitamin A- transporting protein complex occurring in human serum. J. biol. Chem., **246**: 34-43.

PETERSON, P.A. & BERGGÅRD, I. (1971) Isolation and properties of a human retinol-transporting protein. J. biol. Chem., **246**: 25-33.

PETERSON, P.A., EVRIN, P.-E., & BERGGÅRD, I. (1969) Differentiation of glomerular, tubular, and normal proteinuria: determinations of urinary excretion of $\beta$-2-microglobulin, albumin, and total protein. J. clin. Invest., **48**: 1189-1198.

PHILIPP, R. (1985) Cadmium-risk assessment of an exposed residential population: a review. J. R. Soc. Med., **78**: 328-333.

PIOTROWSKI, J.K., TROJANOWSKA, B., WISNIEWSKA-KNYPL, J.M., & BOLANOWSKI, W. (1974) Mercury binding in the kidney and liver of rats repeatedly exposed to mercuric chloride: induction of metallothionein by mercury and cadmium. Toxicol. appl. Pharmacol., **27**: 11-19.

PISCATOR, M. (1962) Proteinuria in chronic cadmium poisoning. II. The applicability of quantitative and qualitative methods of protein determination for the demonstration of cadmium proteinuria. Arch. environ. Health, **5**: 325-332.

PISCATOR, M. (1964) [On cadmium in normal kidneys together with a report on the isolation of metallothionein from livers of cadmium-exposed rabbits.] Nord. Hyg. Tidskr., **45**: 76-82 (in Swedish).

PISCATOR, M. (1966a) Proteinuria in chronic cadmium poisoning. III. Electrophoretic and immunoelectrophoretic studies on urinary proteins from cadmium workers, with special reference to the excretion of low molecular weight proteins. Arch. environ. Health, **12**: 335-344.

PISCATOR, M. (1966b) Proteinuria in chronic cadmium poisoning. IV. Gel filtration and ion exchange chromatography of urinary proteins from cadmium workers. Arch. environ. Health, **12**: 345-359.

PISCATOR, M. (1971) Data on interferences in cadmium analysis. In: Friberg, L., Piscator, M., & Nordberg, G., ed. Cadmium in the environment, Cleveland, Ohio, CRC Press, p. 17.

PISCATOR, M. & TSUCHIYA, K. (1971) In: Friberg, L., Piscator, M., & Nordberg, G., ed. Cadmium in the environment, Cleveland, Ohio, Chemical Rubber Company Press, pp. 112-114.

PISCATOR, M. & VOUK, V.B. (1979) Sampling and analytical methods. In: Friberg, L., Nordberg, G.F., & Vouk, V.B., ed. Handbook on the toxicology of metals, Amsterdam, Oxford, New York, Elsevier Science Publishers, pp. 33-46.

PISCATOR, M., KJELLSTRÖM, T., & LIND, B. (1976) Contamination of cigarettes and pipe tobacco by cadmium-oxide dust. Lancet, **2**: 587.

PISCATOR, M., BJORK, L., & NORDBERG, M. (1981) Beta-2-microglobulin levels in serum and urine of cadmium exposed rabbits. Acta. pharmacol. toxicol., **49**: 1-7.

PLEBAN, P.A., KERKAY, I., & PEARSON, K.H. (1981) Polarized Zeeman-effect flameless atomic absorption spectrometry of cadmium, copper, lead, and manganese in human kidney cortex. Clin. Chem., **27**: 68-72.

POTTS, C.L. (1965) Cadmium proteinuria - the health of battery workers exposed to cadmium oxide dust. Ann. occup. Hyg., **8**: 55-61.

PRIGGE, E. (1978) Early signs of oral and inhalative cadmium uptake in rats. Arch. Toxicol., **40**: 231-247.

PRIGGE, E., BAUMERT, H.P., & MUHLE, H. (1977) Effects of dietary and inhalative cadmium on haemoglobin and haematocrit in rats. Bull. environ. Contam. Toxicol., **17**: 585-590.

PRINCI, F. (1947) A study of industrial exposure to cadmium. J. ind. Hyg. Toxicol., **29**: 315-320.

PRINCI, F. & GEEVER, E.F. (1950) Prolonged inhalation of cadmium. Arch. ind. Hyg. occup. Med., **1**: 651-661.

PROCKOP, D.J. & DAVIDSON, W.D. (1964) A study of urinary and serum lysozyme in patients with renal disease. New Engl. J. Med., **270**: 269-274.

PRODAN, L. (1932) Cadmium poisoning. II. Experimental cadmium poisoning. J. ind. Hyg. Toxicol., **14**: 174-196.

PULIDO, P., KAGI, J.H., & VALLEE, B.L. (1966) Isolation and some properties of human metallothionein. Biochemistry, **5**: 1768-1777.

RAHOLA, T., AARAN, R.-K., & MIETTINEN, J.K. (1972) Half-time studies of mercury and cadmium by whole-body counting. In: Assessment of radioactive contamination in man, New York, International Atomic Energy Agency Unipublishers, pp. 553-562 (IAEA Proceedings Series No. Sm-150/13).

RAMAYA, L.K. & POMERANTZEVA, M.O. (1977) [Investigation of cadmium chloride mutagenic effect on germ cells of male mice.] Genetika, **13**: 59-63 (in Russian).

RAMEL, C. & FRIBERG, L. (1971) Carcinogenic and mutagenic effects. In: Friberg, L., Piscator, M., & Nordberg, G.F., ed. Cadmium in the environment, Cleveland, Ohio, CRC Press, pp. 109-110.

RAUHUT, A. (1980) Survey of industrial emission of cadmium in the European Economic Community: Final Report, Brussels, Commission of the European Communities, Vol. 2, pp. 1-44.

REDDY, C.S., SVOBODA, D., AZARNOFF, D., & DEWAR, R. (1973) Cadmium-induced Leydig cell tumors of rat testis: morphologic and histochemical study. J. Natl Cancer Inst., **51**: 891-903.

RICHARDSON, M.E. & FOX, M.R.S. (1974) Dietary cadmium and enteropathy in the Japanese quail. Lab. Invest., **31**: 722-731.

RIVM (1988) In: Ros, J.P.M. & Sloof, W., ed. Integrated criteria document cadmium, Bilthoven, The Netherlands, National Institute of Public Health and Environmental Protection (RIVM-Report No. 758476004).

ROBBINS, S.L., COTRAN, R.S., & KUMAR, U. (1984) Pathological basis of disease, 3rd ed., Philadelphia, Pennsylvania, W.B. Saunders & Co., pp. 1327-1328.

ROELS, H., HUBERMONT, G., BUCHET, J.P., & LAUWERYS, R. (1978) Placental transfer of lead, mercury, cadmium and carbon monoxide in women. III. Factors influencing the accumulation of heavy metals in the aplacenta and the relationship between metal concentration in the placenta and in maternal and cord blood. Environ. Res., **16**: 236-245.

ROELS, H., LAUWERYS, R.R., BUCHET, J.P., & BERNARD, A. (1981a) Environmental exposure to cadmium and renal function of aged women in three areas of Belgium. Environ. Res., **24**: 117-130.

ROELS, H., LAUWERYS, R.R., BUCHET, J.P., BERNARD, A., CHETTLE, D.R., HARVEY, T.C., & AL-HADDAD, J.K. (1981b) *In vivo* measurement of liver and kidney cadmium in workers exposed to this metal: its significance with respect to cadmium in blood and urine. Environ. Res., **26**: 217-240.

ROELS, H., DJUBGANG, J., BUCHET, J.P., BERNARD, A., & LAUWERYS, R.R. (1982) Evolution of cadmium-induced renal dysfunction in workers removed from exposure. Scand. J. Work Environ. Health, **8**: 191-200.

ROELS, H., LAUWERYS, R.R., & DARDENNE, A.N. (1983a) The critical level of cadmium in human renal cortex: a reevaluation. Toxicol. Lett., **15**: 357-360.

ROELS, H., LAUWERYS, R.R., BUCHET, J.P., BERNARD, A., GARVEY, J.S., & LINTON, H.J. (1983b) Significance of urinary metallothionein in workers exposed to cadmium. Int. Arch. occup. environ. Health, **52**: 159-166.

ROELS, H., LAUWERYS, R.R., BUCHET, J.P., BERNARD, A.M., VOS, A., & OVERSTEYNS, M. (1989) Health significance of cadmium induced renal dysfunction: a five year follow up. Br. J. ind. Med., **46**: 755-764.

ROHR, G. & BAUCHINGER, M. (1976) Chromosome analyses in cell cultures of the Chinese hamster after application of cadmium sulfate. Mutat. Res., **40**: 125-130.

SAITO, H. (1987) [Follow-up study on health status on residents in a cadmium polluted area (Sasu, Izahara town of Nagasaki Prefecture) after improvement of cadmium-polluted lands], Tokyo, Japan Public Health Association, pp. 215-216 (Kankyo Hoken Report No. 53) (in Japanese).

SAITO, H., SHIOJI, R., FURUKAWA, Y., NAGAI, K., ARIKAWA, T., SAITO, T., SASAKI, Y., FURUYAMA, T., & YOSHINAGA, K. (1977) Cadmium-induced proximal tubular dysfunction in a cadmium-polluted area. Nephrology, **6**: 1-12.

SAITO, H., NAKANO, A., TOHYAMA, C., MITANE, Y., SUGIHIRA, N., & WAKISAKA, I. (1983) [Report on a health examination of inhabitants of areas with cadmium pollution of soil], Tokyo, Japan Public Health Association, pp. 91-95 (Kankyo Hoken Report No. 40) (in Japanese).

SALMELA, S. & VUORI, E. (1979) Contamination with cadmium from micropipette tips. Talanta, **26**: 175-176.

SALTZMAN, B.E. (1953) Colorimetric microdetermination of cadmium with dithizone. Anal. Chem., **25**: 493-496.

SAMARAWICKRAMA, G.P. & WEBB, M. (1979) Acute effects of cadmium on the pregnant rat and embryo-fetal development. Environ. Health Perspect., **28**: 245-249.

SANDERS, C.L. & MAHAFFEY, J.A. (1984) Carcinogenicity of single and multiple intratracheal instillations of cadmium oxide in the rat. Environ. Res., **33**: 227-233.

SANGSTER, B., DE GROOT, G., LOEBER, J.G., DERKS, H.J.G.M., KRAJNC, E.I., & SAVELKOUL, T.J.F. (1984) Urinary excretion of cadmium, protein, beta-2-microglobulin and glucose in individuals living in a cadmium-polluted area. Hum. Toxicol., **3**: 7-21.

SCHARPF, L.G., Jr, HILL, I.D., WRIGHT, P.L., PLANK, J.B., KEPLINGER, M.L., & CALANDRA, J.C. (1972) Effect of sodium nitrilotriacetate on toxicity, teratogenicity, and tissue distribution of cadmium. Nature (Lond.), **239**: 231-234.

SCHERER, G. & BARKEMEYER, H. (1983) Cadmium concentrations in tobacco and tobacco smoke. Ecotoxicol. environ. Saf., **7**: 71-78.

SCHMID, B.P., KAO, J., & GOULDING, E. (1985) Evidence for reopening of the cranial tube in mouse embryo treated with cadmium chloride. Experientia (Basel), **41**: 271-272.

SCHROEDER, H.A. (1964) Cadmium hypertension in rats. Am. J. Physiol., **207**: 62-66.

SCHROEDER, H.A. (1965) Cadmium as a factor in hypertension. J. chron. Dis., **18**: 647-656.

SCHROEDER, H.A. (1967) Cadmium, chromium, and cardiovascular disease. Circulation, **35**: 570.

SCHROEDER, H.A. & BALASSA, J.J. (1961) Abnormal trace metals in man: cadmium. J. chron. Dis., **14**: 236-258.

SCHROEDER, H.A. & MITCHENER, M. (1971) Toxic effects of trace elements on the reproduction of mice and rats. Arch. environ. Health, **23**: 102-106.

SCHROEDER, H.A. & VINTON, W.H., Jr (1962) Hypertension induced in rats by small doses of cadmium. Am. J. Physiol., **202**: 515-518.

SCHROEDER, H.A., BALASSA, J.J., & VINTON, W.H., Jr (1964) Chromium, lead, cadmium, nickel, and titanium in mice: effect on mortality, tumours, and tissue levels. J. Nutr., **83**: 239-250.

SCHROEDER, H.A., BALASSA, J.J., & VINTON, W.H., Jr (1965) Chromium, cadmium, and lead in rats: effects on life span, tumours, and tissue levels. J. Nutr., **86**: 51-66.

SCOTT, R., MILLS, E.A., FELL, G.S., HUSAIN, F.E.R., YATES, A.J., PATERSON, P.J., MCKIRDY, A., OTTOWAY, J.M., FITZGERALD-FINCH, O.P., & LAMONT, A. (1976) Clinical and biochemical abnormalities in coppersmiths exposed to cadmium. Lancet, **2**: 396-398.

SCOTT, R., AUGHEY, E., & SINCLAIR, J. (1977) Histological and ultrastructural changes in rat kidney following cadmium injection. Urol. Res., **5**: 15-20.

SCOTT, R., PATERSON, P.J., BURNS, R., OTTOWAY, J.M., HUSAIN, F.E.R., FELL, G.S., DUMBUYA, S., & IGBAL, M. (1978) Hypercalciuria related to cadmium exposure. Urology, **11**: 462-465.

SCOTT, R., PATERSON, P.J., BURNS, R., FELL, G.S., & OTTOWAY, J.M. (1979) The effects of treatment on the hypercalciuria of chronic cadmium poisoning. Urol. Res., **7**: 285-289.

SCOTT, R., HAYWOOD, J.K., BODDY, K., WILLIAMS, E.D., HARVEY, M., & PATERSON, P.J. (1980) Whole body calcium deficit in cadmium-exposed workers with hypercalciuria. Urology, **15**: 356-359.

SHAIKH, Z.A. & HIRAYAMA, K. (1979) Metallothionein in the extracellular fluids as an index of cadmium toxicity. Environ. Health Perspect., **28**: 267-272.

SHAIKH, Z.A. & SMITH, J.C. (1980) Metabolism of orally ingested cadmium in humans. In: Holmstedt, B., Lauwerys, R., Mercier, M., & Roberfroid, M., ed. Mechanisms of toxicity and hazard evaluation, Amsterdam, Elsevier/North Holland Biomedical Press, pp. 569-574.

SHARMA, R.P., KJELLSTRÖM, T., & MCKENZIE, J.M. (1983) Cadmium in blood and urine among smokers and non-smokers with high cadmium intake via food. Toxicology, **29**: 163-171.

SHARRETT, A.R., CARTER, A.P., ORHEIM, R.M., & FEINLEIB, M. (1982) Daily intake of lead, cadmium, copper, and zinc from drinking-water: the Seattle study of trace metal exposure. Environ. Res., **28**: 456-475.

SHERLOCK, J. & WALTERS, B. (1983) Dietary intake of heavy metals and its estimation. Chem. Ind., **13**: 505-508.

SHERLOCK, J., SMART, G.A., & WALTERS, B. (1983) Dietary surveys on a population at Shipham, Somerset, United Kingdom. Sci. total Environ., **29**: 121-142.

SHIGEMATSU, I. & KAWAGUCHI, T. (1978) Environmental pollution and health effects, Yamagata Prefecture. In: Tsuchiya, K., ed. Cadmium studies in Japan: a review, Amsterdam, Oxford, New York, Elsevier Science Publishers, pp. 199-211.

SHIGEMATSU, I., KAWAGUCHI, T., & YANAGAWA, H. (1978) Reanalysis of the results of health examinations conducted on the general population in the cadmium polluted areas of Japan. In: Tsuchiya, K. ed. Cadmium studies in Japan: a review, Amsterdam, Oxford, New York, Elsevier Science Publishers, pp. 253-264.

SHIGEMATSU, I., YANAGAWA, H., & KAWAGUCHI, T. (1975) [Record filling of health examinations for inhabitants in the Cd-polluted areas (2nd and 3rd reports)], Tokyo, Japan Public Health Association, pp. 72-101 (Kankyo Hoken Report No. 36) (in Japanese).

SHIGEMATSU, I., MINOWA, M., YOSHIDA, T., & MIYAMOTO, K. (1979) Recent results of health examinations on the general population in cadmium-polluted and control areas in Japan. Environ. Health Perspect., **28**: 205-210.

SHIGEMATSU, I., MINOWA, M., YOSHIDA, T., & MIYAMOTO, K. (1980) Epidemiological aspects of environmental cadmium and its effects in Japan. In: Proceedings of the 2nd International Cadmium Conference, Cannes, 6-8 February, 1979, London, Cadmium Association, pp. 111-117.

SHIGEMATSU, I., KITAMURA, S., TAKEUCHI, J., MINOWA, M., NAGAI, M., USUI, T., & FUKUSHIMA, M. (1981) A retrospective mortality study on cadmium-exposed populations in Japan. In: Recent studies on health effects of cadmium in Japan, Tokyo, The Japan Cadmium Research Committee, Japan Environment Agency, pp. 303-317.

SHIGEMATSU, I., MITAMURA, S., TAKEUCHI, J., MINOWA, M., NAGAI, M., & FUKUSHIMA, M. (1982) A retrospective mortality study on cadmium-exposed populations in Japan. In: Proceedings of the 3rd International Cadmium Conference, Miami, Florida, 3-5 February, 1981, London, Cadmium Association, pp. 115-118.

SHIGEMATSU, I., MINOWA, M., NAGAI, M., OHMURA, T., & TAKEUCHI, K. (1983) A retrospective mortality study on cadmium-exposed populations in Japan (supplement): mortalities by level of pollution in Toyama Prefecture. In: The influence of nutritional factors on cadmium administered to monkeys, Tokyo, Japan Environment Agency, pp. 158-174.

SHILLER, A.M. & BOYLE, E.A. (1987) Variability of dissolved trace metals in the Mississippi River. Geochim. Cosmochim. Acta., **51**: 3273-3277.

SHIRAISHI, Y. (1975) Cytogenetic studies in 12 patients with Itai-itai disease. Humangenetik, **27**: 31-44.

SHIRAISHI, Y., KURAHASHI, H., & YOSHIDA, T.H. (1972) Chromosomal aberrations in cultured human leukocytes induced cadmium sulfide. Proc. Jpn. Acad., **48**(21): 133-137.

SHIROISHI, K., ANAYAMA, M., TANII, M., FUKUYAMA, Y., & KUBOTA, K. (1972) [Urinary findings of different age groups in the inhabitants of Itai-itai disease districts.] Med. Biol., **85**: 263-267 (in Japanese).

SHIROISHI, K., KJELLSTROM, T., ANAYAMA, M., SHIMADA, T., IWATA, T., NISHINO, H., MATSUNAGA, A., WATANABE, M., & KUBOTA, K. (1975) [Studies on the urinary β-2-microglobulin level of the inhabitants in Cd-polluted districts.] Jpn J. Hyg., **30**: 71 (in Japanese).

SHIROISHI, K., KJELLSTRÖM, T., KUBOTA, K., EVRIN, P.E., ANAYAMA, M., VESTERBERG, O., SHIMADA, T., PISCATOR, M., IWATA, T., & NISHINO, H. (1977) Urine analysis for detection of cadmium-induced renal changes, with special reference to β-2-microglobulin. Environ. Res., **13**: 407-424.

SIMPSON, W.R. (1981) A critical review of cadmium in the marine environment. Prog. Oceanogr., **10**: 1-70.

SKERFVING, S., CHRISTOFFERSON, J.O., SHUTZ, A., WELINDER, H., SPANG, G., AHLGREN, L., & MATTSON, S. (1987) Biological monitoring, by *in vivo* XRF measurements, of occupational exposure to lead, cadmium and mercury. Biol. Trace Elem. Res., **13**: 241-251.

SKOG, E. & WAHLBERG, J.E. (1964) A comparative investigation of percutaneous absorption of metal compounds in guinea-pig by means of the radioactive isotopes, $^{51}$Cr, $^{58}$Co, $^{65}$Zn, $^{110}$Ag, $^{115}$Cd, $^{203}$Hg. J. invest. Dermatol., **43**: 187-192.

SMITH, J.P., SMITH, J.C., & MCCALL, A.J. (1960) Chronic poisoning from cadmium fume. J. Pathol. Bacteriol., **80**: 287-296.

SMITH, T. (1982) Emission inventory case studies - policy aspects. J. Air Pollut. Control Assoc., **32**(10): 1017-1020.

SMITH, T., PETTY, T.L., READING, J.C., & LAKSHMINARAYAN, S. (1976) Pulmonary effects of chronic exposure to airborne cadmium. Am. Rev. respir. Dis., **114**: 161-169.

SMITH, T., ANDERSON, R.J., & READING, J.C. (1980) Chronic cadmium exposures associated with kidney function effects. Am. J. ind. Med., **1**: 319-337.

SNIDER, G.L., HAYES, J.A., KORTHY, A.L., & LEWIS, G.P. (1973) Centrilobular emphysema experimentally induced by cadmium chloride aerosol. Am. Rev. respir. Dis., **108**: 40-48.

SONAWANE, B.R., NORDBERG, M., NORDBERG, G.F., & LUCIER, G.W. (1975) Placental transfer of cadmium in rats: influence of dose and gestational age. Environ. Health Perspect., **12**: 97-102.

SORAHAN, T. (1987) Mortality from lung cancer among a cohort of nickel cadmium battery workers: 1946-84. Br. J. ind. Med., **44**: 803-809.

SORAHAN, T. & WATERHOUSE, J.A.H. (1983) Mortality study of nickel-cadmium battery workers by the method of regression models in life tables. Br. J. ind. Med., **40**: 293-300.

SPENCER, H., ASMUSSEN, C.R., HOLTZMAN, R.B., & KRAMER, L. (1979) Metabolic balances of cadmium, copper, manganese, and zinc in man. Am. J. clin. Nutr., **32**: 1867-1875.

SPORN, A., DINU, I., & STOENESCU, L. (1970) Influence of cadmium administration on carbohydrate and cellular energetic metabolism in the rat liver. Rev. roum. Biochim., **7**: 299-305.

SQUIBB, K.S., PRITCHARD, J.B., & FOWLER, B.A. (1982) Renal metabolism and toxicity of metallothionein. In: Foulkes, E.C., ed. Biological roles of metallothionein, Amsterdam, Oxford, New York, Elsevier Science Publishers, pp. 181-190.

SQUIBB, K.S., PRITCHARD, J.B., & FOWLER, B.A. (1984) Cadmium-metallothionein nephropathy: relationships between ultrastructural/biochemical alterations and intracellular cadmium binding. J. Pharmacol. exp. Ther., **229**(1): 311-321.

SQUIRE, J.R., HARDWICKE, J., & SOOTHILL, J.F. (1962) Proteinuria. In: Black, D.A.K., ed. Renal disease, Oxford, Blackwell Scientific Publishers, pp. 218-231.

STACEY, N.H., CANTILENA, L.R., & KLAASSEN, C.D. (1980) Cadmium toxicity and lipid peroxidation in isolated rat hepatocytes. Toxicol. appl. Pharmacol., **53**: 470-480.

STAESSEN, J., BULPITT, C.J., ROELS, H., BERNARD, A., FAGARD, R., JOOSSENS, J.V., LAUWERYS, R.R., LIJNEN, P., & AMERY, A. (1984) Urinary cadmium and lead and their relationship to blood pressure in a population with low average exposure. Br. J. ind. Med., **41**: 241-248.

STANESCU, D., VEITNER, C., FRANS, A., CONCETTE, L., ROELS, H., LAUWERYS, R.R., & BRASSEUR, L. (1977) Effects on lung of chronic occupational exposure to cadmium. Scand. J. respir. Dis., **58**: 289-303.

STEWART, M. & HUGHES, E.G. (1981) Urinary beta-2-microglobin in the biological monitoring of cadmium workers. Br. J. Ind. Med., **38**: 170-174.

STOEPPLER, M. & BRANDT, K. (1978) Determination of lead and cadmium in whole blood by electrothermal atomic absorption spectroscopy. In: Doss, M., ed. Proceedings of an International Symposium on Clinical Biochemistry on Diagnosis and Therapy of Porphyrias and Lead Intoxication, New York, Springer-Verlag, pp. 185-187.

STOEPPLER, M. & BRANDT, K. (1980) Contributions to automated trace analysis. V. Determination of cadmium in whole blood and urine by electrothermal atomic absorption spectrophotometry. Fresenius Z. anal. Chem., **300**: 372-380.

STOEPPLER, M., VALENTA, P., & NURNBERG, H.W. (1979) Applications of independent methods and standard materials: an effective approach to reliable trace and ultra trace analysis of metals and metalloids in environmental and biological materials. Fresenius Z. anal. Chem., **297**: 22-34.

STONEBURNER, D.L. (1978) Heavy metals in tissues of stranded short-finned pilot whales. Sci. total Environ., **9**: 293-297.

STOWE, H.D., WILSON, M., & GOYER, R.A. (1972) Clinical and morphological effects of oral cadmium toxicity in rabbits. Arch. Pathol., **94**: 389-405.

STRAUSS, R.H., PALMER, K.C., & HAYES, J.A. (1976) Acute lung injury induced by cadmium aerosol. I. Evolution of alveolar cell damage. Am. J. Pathol., **84**: 561-568.

SUBRAMANIAN, K.S. & MERANGER, J.C. (1981) A rapid electrothermal atomic absorption spectrophotometric method for cadmium and lead in human whole blood. Clin. Chem., **27**: 1866-1871.

SUGAWARA, C. & SUGAWARA, N. (1974) Cadmium toxicity for rat intestine, especially on the absorption of calcium and phosphorus. Jpn. J. Hyg., **28**: 511-516.

SULLIVAN, M.F., HARDY, J.T., MILLER, B.M., BUSCHBOM, R.L., & SIEWICKI, T.C. (1984) Absorption and distribution of cadmium in mice fed diets containing either inorganic or oyster-incorporated cadmium. Toxicol. appl. Pharmacol., **72**: 210-217.

SUMINO, K., HAYAKAWA, K., SHIBATA, T., & KITAMURA, S. (1975) Heavy metals in normal Japanese tissues. Arch. environ. Health, **30**: 487-494.

SUZUKI, S. & LU, C.C. (1976) A balance study of cadmium: an estimation of daily input, output, and retained amount in two subjects. Ind. Health, **14**: 53-65.

SUZUKI, S., SUZUKI, T., & ASHIZAWA, M. (1965) Proteinuria due to inhalation of cadmium stearate dust. Ind. Health, **3**: 73-85.

SUZUKI, S., TAGUCHI, T., & YOKOHASHI, G. (1969) Dietary factors influencing upon the retention rate of orally-administered $^{115}CdCl_2$ in mice, with special reference to calcium and protein concentration in diet. Ind. Health, **7**: 155-162.

SUZUKI, Y. (1980) Cadmium metabolism and toxicity in rats after long-term subcutaneous administration. J. Toxicol. environ. Health, **6**: 469-482.

SUZUKI, Y. (1982) Metal-binding properties of metallothionein in extracellular fluids and its role in cadmium-exposed rats. In: Foulkes, E.C., ed. Biological roles of metallothionein, Amsterdam, Oxford, New York, Elsevier Science Publishers, pp. 25-67.

SVARTENGREN, M., ELINDER, C.G., FRIBERG, L., & LIND, B. (1986) Distribution and concentration of cadmium in human kidney. Environ. Res., **39**: 1-7.

SVENSSON, B.-G., BJORNHAM, A., SCHUTZ, A., LETTEVALL, U., NILSSON, A., & SKERFVING, S. (1987) Acidic deposition and human exposure to toxic metals. Sci. total Environ., 67: 101-115.

TAKABATAKE, E. (1978a) Environmental pollution and health effects: Oita Prefecture. In: Tsuchiya, K., ed. Cadmium studies in Japan: a review, Amsterdam, Oxford, New York, Elsevier Science Publishers, pp. 215-218.

TAKABATAKE, E. (1978b) Nagasaki Prefecture. In: Tsuchiya, K., ed. Cadmium studies in Japan: a review, Amsterdam, Oxford, New York, Elsevier Science Publishers, pp. 181-192.

TAKASHIMA, M., MORIWAKI, S., & ITOKAWA, Y. (1980) Osteomalacic change induced by long-term administration of cadmium to rats. Toxicol. appl. Pharmacol., 54: 223-228.

TAKEBAYASHI, S. (1980a) First autopsy case, suspicious of cadmium intoxication, from the cadmium-polluted area in Tsushima, Nagasaki Prefecture. In: Shigematsu, I. & Nomiyama, K., ed. Cadmium-induced osteopathy, Tokyo, Japan Public Health Association, pp. 124-138.

TAKEBAYASHI, S. (1980b) [Clinical and pathological findings of a case with osteomalacia in a cadmium-polluted area (Tsushima Island) (1st autopsied case)], Tokyo, Japan Public Health Association, pp. 259-262 (Kankyo Hoken Report No. 46) (in Japanese).

TAKEBAYASHI, S. (1983a) [Clinical and pathological findings of a case with osteomalacia in a cadmium-polluted area (Tsushima Island) (2nd autopsied case)], Tokyo, Japan Public Health Association, pp. 207-211 (Kankyo Hoken Report No. 49) (in Japanese).

TAKEBAYASHI, S. (1983b) [Clinical and pathological findings of a case with diabetic nephropathy in a cadmium-polluted area (Tsushima Island) (3rd autopsied case)], Tokyo, Japan Public Health Association, pp. 212-217 (Kankyo Hoken Report No. 49) (in Japanese).

TAKEBAYASHI, S. (1984) [Clinical and pathological findings of a case with osteomalacia in a cadmium-polluted area (Tsushima Island) (4th autopsied case)], Tokyo, Japan Public Health Association, pp. 238-249 (Kankyo Hoken Report No. 50) (in Japanese).

TAKEBAYASHI, S., SATO, T., HARADA, K., & YOSHIMURA, S. (1985) [Clinical and pathological findings of a case with osteomalacia in a cadmium-polluted area (Tsushima Island) (5th autopsied case)], Tokyo, Japan Public Health Association, pp. 245-251 (Kankyo Hoken Report No. 51) (in Japanese).

TAKEBAYASHI, S., SATO, T., HARADA, K., HIRAI, Y., & YOSHIMURA, S. (1987a) [Clinical and pathological findings of a case with osteomalacia in a cadmium-polluted area (Tsushima Island) (6th autopsied case)], Tokyo, Japan Public Health Association, pp. 294-305 (Kankyo Hoken Report No. 53) (in Japanese).

TAKEBAYASHI, S., SATO, T., HARADA, K., HIRAI, Y., & YOSHIMURA, S. (1987b) [Clinical and pathological findings of a case with osteomalacia in a cadmium-polluted area (Tsushima Island) (7th autopsied case)], Tokyo, Japan Public Health Association, pp. 306-319 (Kankyo Hoken Report No. 53) (in Japanese).

TAKEBAYASHI, S., HIRAI, Y., & YOSHIMURA, S. (1988a) [Clinical and pathological findings of a case with osteomalacia in a cadmium-polluted area (Tsushima Island) (8th autopsied case)], Tokyo, Japan Public Health Association, pp. 187-192 (Kankyo Hoken Report No. 54) (in Japanese).

TAKEBAYASHI, S., HIRAI, Y., & YOSHIMURA, S. (1988b) [Clinical and pathological findings of a case with osteomalacia in a cadmium-polluted area (Tsushima Island) (9th autopsied case)], Tokyo, Japan Public Health Association, pp. 193-199 (Kankyo Hoken Report No. 54) (in Japanese).

TAKEBAYASHI, S., HIRAI, Y., & YOSHIMURA, S. (1988c) [Clinical and pathological findings of a case with osteomalacia in a cadmium-polluted area (Tsushima Island) (10th autopsied case)], Tokyo, Japan Public Health Association, pp. 200-206 (Kankyo Hoken Report No. 54) (in Japanese).

TAKEBAYASHI, S., HARADA, K., & YOSHIMURA, S. (1988d) [Clinical and pathological findings of a case with osteomalacia in a cadmium-polluted area (Tsushima Island) (11th autopsied case)], Tokyo, Japan Public Health Association, pp. 207-213 (Kankyo Hoken Report No. 54) (in Japanese).

TAKENAKA, S., OLDIGES, H., KOENIG, H., HOCHRAINER, D., & OBERDOERSTER, G. (1983) Carcinogenicity of cadmium aerosols in Wistar rats. J. Natl Cancer Inst., **70**: 367-371.

TANAKA, K. (1982) Effects of hepatic disorders on the fate of cadmium in rats. In: Foulkes, E.C., ed. Biological roles of metallothionein, Amsterdam, Oxford, New York, Elsevier Science Publishers, pp. 237-249.

TANAKA, K., MIN, K.-S., ONOSAKA, S., FUKUHARA, C., & UEDA, M. (1985) The origin of metallothionein in red blood cells. Toxicol. appl. Pharmacol., **78**: 63-66.

TARASENKO, N.YU. & VOROBJEVA, R.S. (1973) [Hygienic problems connected with the use of cadmium.] Vestn. Akad. Med. Nauk. SSR, **28**: 37-43 (in Russian with English abstract).

TARASENKO, N.YU., VOROBJEVA, R.S., SPIRIDINOVA, V.S., & SABALINA, L.P. (1974) Experimental investigation of toxicity of cadmium and zinc caprylates. J. Hyg. Epidemiol. Microbiol. Immunol., **18**: 144-153.

TARASENKO, N.YU., VOROBJEVA, R.S., SABALINA, L.P., & CVETKOVA, R.P. (1975) [Cadmium in the environment and its effect on calcium metabolism.] Gig. i Sanit., **9**: 22-25 (in Russian).

TASK GROUP ON LUNG DYNAMICS (1966) Deposition and retention models for internal dosimetry of the human respiratory tract. Health Phys., **12**: 173-208.

TASK GROUP ON METAL INTERACTIONS (1978) Factors influencing metabolism and toxicity of metals: a consensus report. Environ. Health Perspect., **25**: 3-42.

TASK GROUP ON METAL TOXICITY (1976) Conceptual considerations: critical organ, critical concentration in cells and organs, critical effect, subcritical effect, dose-effect and dose-response relationships. In: Nordberg, G.F., ed. Effects and dose-response relationships of toxic metals, Amsterdam, Oxford, New York, Elsevier Science Publishers, pp. 10-13.

TATI, M., KATAGIRI, Y., & KAWAI, M. (1976) Urinary and faecal excretion of cadmium in normal Japanese: an approach to non-toxic levels of cadmium. In: Nordberg, G.F., ed. Effects and dose-response relationships of toxic metals, Amsterdam, Oxford, New York, Elsevier Science Publishers, pp. 331-341.

TAYLOR, B.A., HEINIGER, H.J., & MEIER, H. (1973) Genetic analysis of resistance to cadmium-induced testicular damage in mice (37380). Proc. Soc. Exp. Biol. Med., **143**: 629-633.

TECULESCU, D.B. & STANESCU, D.C. (1970) Pulmonary function in workers with chronic exposure to cadmium oxide fumes. Int. Arch. Arbeitsmed., **26**: 335-345.

TEMPLETON, D.M., BANERJEE, D., & CHERIAN, M.G. (1985) Metallothionein synthesis and localization in relation to metal storage in rat liver during gestation. Can. J. Biochem. cell Biol., **63**: 16-22.

TENNANT, C.J. (1984) Cadmium in the environment and high risk population groups, Birmingham, University of Aston (Ph.D. Thesis).

TERHAAR, C.J., VIS, E., ROUDABUSH, R.L., & FASSETT, D.W. (1965) Protective effects of low doses of cadmium chloride against subsequent high oral doses in the rat. Toxicol. appl. Pharmacol., **7**: 500-505.

THIEME, R., SCHRAMEL, P., & KURZ, E. (1977) [Trace element concentration in the human placenta in a heavily-contaminated environment.] Geburtshilfe Frauenheilkd., **37**: 756-761 (in German).

THOMAS, B.J., HARVEY, T.C., CHETTLE, D.R., MCLELLAN, J.S., & FREMLIN, J.H. (1979) A transportable system for the measurement of liver cadmium *in vivo*. Phys. Med. Biol., **24**: 432-437.

THORNTON, I. (1988) The Shipham report. An investigation into cadmium contamination and its implications for human health. Metal content of soils and dusts. Sci. total Environ., **75**: 21-39.

THUN, M.T., SCHNORR, T.M., SMITH, A.B., HALPERIN, W.E., & LEMEN, R.A. (1985) Mortality among a cohort of U.S. cadmium production workers: An update. J. Natl Cancer Inst., **74**: 325-333.

THUN, M.J., OSORIO, A.M., SCHOBER, S., HANNON, W.H., & HALPERIN, W. (1989) Nephropathy in cadmium workers - assessment of risk from airborne occupational cadmium exposure. Br. J. ind. Med., **46**: 689-697.

TIPTON, I.H., SCHROEDER, H.A., PERRY, H.M., & COOK, M.J. (1960) Trace elements in human tissues. Part III. Subjects from Africa, the Near and Far East, and Europe. Health Phys., **11**: 403-451.

TJELL, J.C., HANSEN, J.A., CHRISTENSEN T.H., & HOVMAND, M.F. (1981) Prediction of cadmium concentrations in Danish soils. In: L'Hermite, P. & Ott, H., ed. Characterization, treatment and use of sewage sludge, Dordrecht, Reidel Publishing Co., pp. 652-664.

TJELL, J.C., CHRISTENSEN, T.H., & BRO-RASMUSSEN, F. (1983) Cadmium in soil and terrestrial biota, with emphasis on the Danish situation. Ecotoxicol. environ. Saf., 7: 122-140.

TOHYAMA, C., SHAIKH, Z.A., ELLIS, K.J., & COHN, S.H. (1981a) Metallothionein excretion in urine upon cadmium exposure: its relationship with liver and kidney cadmium. Toxicology, **20**: 181-191.

TOHYAMA, C., SHAIKH, Z.A., NOGAWA, K., KOBAYASHI, E., & HONDA, R. (1981b) Elevated urinary excretion of metallothionein due to environmental cadmium exposure. Toxicology, **22**: 289-297.

TOPPING, M.D., FORSTER, H.W., DOLMAN, C., LUCZYNSKA, C.M., & BERNARD, A.M. (1986) Measurement of urinary retinol-binding protein by enzyme-linked immunosorbent assay and its application to detection of tubular proteinuria. Clin. Chem., **32**: 1863-1866.

TSUCHIYA, K. (1967) Proteinuria of workers exposed to cadmium fume. The relationship to concentration in the working environment. Arch. environ. Health, **14**: 875-880.

TSUCHIYA, K. (1969) Causation of Ouch-ouch disease (Itai-itai byo): an introduction review. Part 1. Nature of the disease. Keio J. Med., **18**: 181-194.

TSUCHIYA, K. (1976) Proteinuria of cadmium workers. J. occup. Med., **18**: 463-466.

TSUCHIYA, K., ed. (1978) Cadmium studies in Japan: a review, Amsterdam, Oxford, New York, Elsevier Science Publishers, 376 pp.

TSUCHIYA, K. & IWAO, S. (1978) Interrelationships among zinc, copper, lead, and cadmium in food, faeces, and organs of humans. Environ. Health Perspect., **25**: 119-124.

TSUCHIYA, K. & NAKAMURA, K.I. (1978) Environmental pollution and health effects: Hyogo prefecture. In: Tsuchiya, K., ed. Cadmium studies in Japan: a review, Amsterdam, Oxford, New York, Elsevier Science Publishers, pp. 146-167.

TSUCHIYA, K. & SUGITA, M. (1971) A mathematical model for deriving the biological half-life of a chemical. Nord. Hyg. Tidskr., **53**: 105-110.

TSUCHIYA, K., SEKI, Y., & SUGITA, M. (1976) Cadmium concentrations in the organs and tissues of cadavers from accidental deaths. Keio J. Med., **25**: 83-90.

TSUCHIYA, K., IWAO, S., SUGITA, M., & SAKURAI, H. (1979) Increased urinary $\beta$-2-microglobulin in cadmium exposure: dose-effect relationship and biological significance of $\beta$-2-microglobulin. Environ. Health Perspect., **28**: 147-153.

URUNO, K., OTOYAMA, S., NAKANO, N., & ITAGAKI, Z. (1975) [Results of health examination for inhabitants in Yoshinokawa area polluted with cadmium], Tokyo, Japan Public Health Association, pp. 109-113 (Kankyo Hoken Report No. 36) (in Japanese).

USHIO, F. & DOGUCHI, M. (1977) Dietary intakes of some chlorinated hydrocarbons and heavy metals estimated on the experimentally-prepared diets. Bull. environ. Contam. Toxicol., **17**: 707-711.

US PUBLIC HEALTH SERVICE (1942) Cadmium poisoning. Public Health Rep., **57**: 602-612.

VAHTER, M., ed. (1982) Assessment of human exposure to lead and cadmium through biological monitoring, Stockholm, National Swedish Institute of Environmental Medicine and Department of Environmental Hygiene, Karolinska Institute, 136 pp (Report prepared for the United Nations Environment Programme and the World Health Organization).

VALETAS, P. (1946) "Cadmiose" ou l'intoxication du cadmium, Paris, School of Medicine, 40 pp (Medical Faculty Thesis Report).

VAN AALST, R.M., VAN ARDENNE, R.A.M., DE KRUK, J.F., & LEMS, T. (1983a) Pollution of the North Sea from the atmosphere, Apeldoorn, The Netherlands Organization for Applied Scientific Research (TNO) (Report No. C182/152).

VAN AALST, R.M., DUYZER, J.H., & VELDT, C. (1983b) Atmospheric deposition of lead and cadmium in the southern part of the North Sea. I. Emission and preliminary model calculations, Apeldoorn, The Netherlands Organization for Applied Scientific Research (TNO) (Report No. R83/222).

VANDER, A.J. (1962) Cadmium enhancement of proximal tubular sodium reabsorption. Am. J. Physiol., **203**: 1005-1007.

VANDER MALLIE, R.J. & GARVEY, J.S. (1979) Radioimmunoassay of metallothioneins. J. biol. Chem., **254**: 8416-8421.

VENS, M.D. & LAUWERYS, R.R. (1982) Détermination simultanée du plomb et du cadmium dans le sang et l'urine par le couplage des techniques de chromatographie sur résine échangeuse d'ions et de spectrophotométrie d'absorption atomique. Arch. Mal. prof. Méd. Trav. Sécur. soc., **33**: 97-105.

VESTERBERG, O. & WRANGSKOGH, K. (1978) Determination of cadmium in urine by graphite furnace atomic absorption spectroscopy. Clin. Chem., **24**: 681-685.

VOROBJEVA, R.S. (1957) [Investigations of the nervous system function in workers exposed to cadmium oxide.] Neuropatol. Psikkiatr., **57**: 385-388 (in Russian).

VOROBJEVA, R.S. (1958) [About occupational cadmium oxide poisonings.] J. Hyg. Epidemiol. Microbiol. Immunol., **11**: 244-249 (in German).

VOROBJEVA, R.S. & BUBNOVA, N.I. (1981) [Effects of elemental cadmium and telluric cadmium on organisms.] Gig. Trud. prof. Zabol., **2**: 42-43 (in Russian).

VOROBJEVA, R.S. & EREMEEVA, E.P. (1980) [Cardiovascular function in workers exposed to cadmium.] Gig. i Sanit., **10**: 22-25 (in Russian).

VOROBJEVA, R.S. & SABALINA, L.P. (1975) [Experimental investigations of toxic properties of various cadmium compounds.] Gig. i Sanit., **2**: 102-104 (in Russian).

VOSTAL, J.J. (1976) Dose-response relationship of mercury and cadmium effects on renal cadmium transport. In: Nordberg, G.F., ed. Effects and dose-response relationships of toxic metals, Amsterdam, Oxford, New York, Elsevier Science Publishers, pp. 299-308.

VUORI, E., HUUNAN-SEPPALA, A., KILPIO, J.O., & SALMELA, S.S. (1979) Biologically-active metals in human tissues. II. The effect of age on the concentration of cadmium in aorta, heart, kidney, liver, lung, pancreas, and skeletal muscle. Scand. J. Work Environ. Health, **5**: 16-22.

WAALKES, M.P., POISNER, A.M., WOOD, G.W., & KLAASSEN, C.D. (1984) Metallothionein-like proteins in human placenta and fetal membranes. Toxicol. appl. Pharmacol., **74**: 179-184.

WADE, G.G. (1987) Marked synergism of dimethylnitrosamine carcinogenesis in rats exposed to cadmium. Cancer Res., **47**: 6606-6613.

WALSH, J.J. & BURCH, G.E. (1959) The rate of disappearance from plasma and subsequent distribution of radiocadmium ( 115 m Cd) in normal dogs. J. clin. Med., **54**: 59-65.

WALTERS, B. & SHERLOCK, J. (1981) Studies on the dietary intake of heavy metals. In: Proceedings of the International Conference on Heavy Metals in the Environment, Amsterdam, September 1981, Edinburgh, CEP Consultants Ltd, pp. 506-512.

WATANABE, H. & MURAYAMA, H. (1975) A study on health effect indices concerning population in a cadmium-polluted area. In: International Symposium on Recent Advances in the Assessment of the Health Effects of Environmental Pollution, Luxembourg, December 1975, Luxembourg, Commission of the European Communities, pp. 2185-2195 (Report EURO 5360).

WATANABE, I., KOIZUMI, A., FUJITA, H., KUMAI, M., & IKEDA, M. (1985) Dietary cadmium intakes of farmers in nonpolluted areas in Japan, and the relation with blood cadmium levels. Environ. Res., **37**(1): 33-43.

WATANABE, T., SHIMADA, T., & ENDO, A. (1979) Mutagenic effects of cadmium on mammalian oocyte chromosomes. Mutat. Res., **67**: 349-356.

WEAST, R.C. (1974) Handbook of chemistry and physics, 55th ed., Cleveland, Ohio, CRC Press, 500 pp.

WEBB, M. (1972) Protection by zinc against cadmium toxicity. Biochem. Pharmacol., **21**: 2767-2771.

WEBB, M. & CAIN, K. (1982) Functions of metallothionein. Biochem. Pharmacol., **31**: 137-142.

WEBB, M. & ETIENNE, A.T. (1977) Studies on the toxicity and metabolism of cadmium-thionein. Biochem. Pharmacol., **26**: 25-30.

WESTER, P.O. (1974) Trace element balances in relation to variations in calcium intake. Atherosclerosis, **20**: 207-215.

WHANGER, P.D. (1979) Cadmium effects in rats on tissue iron, selenium, and blood pressure: blood and hair cadmium in some Oregon residents. Environ. Health Perspect., **28**: 115-121.

WHO (1980) Recommended health-based limits in occupational exposure to heavy metals. Report of a study group, Geneva, World Health Organization (Technical Report Series No. 647).

WHO (1984) Guidelines for drinking-water quality: Health criteria and other supporting information, Geneva, World Health Organization, Vol. 2, pp. 84-90.

WHO (1986) Health impact of acidic deposition. Sci. total Environ., **52**: 157-187.

WHO (1987) Air quality guidelines for Europe, Copenhagen, World Health Organization, Regional Office for Europe, pp. 200-209 (European Series, Vol. 23).

WHO (1989) Cadmium. In: Evaluation of certain food additives and contaminants. Thirty-third Report of the Joint FAO/WHO Expert Committee on Food Additives, Geneva, World Health Organization, pp. 28-31 (Technical Report Series No. 776).

WILLIAMS, C.H. & DAVID, D.J. (1973) The effect of superphosphate on the cadmium content of soils and plants. Aust. J. Soil Res., **11**: 43-56.

WILLIAMS, C.H. & DAVID, D.J. (1976) The accumulation in soil of cadmium residues from phosphate for fertilizers and their effect on the cadmium content of plants. Soil Sci., **121**: 86-93.

WILSON, B. (1988a) Investigation of trace metals in the aqueous environment: Final report (January 1986-December 1987), Houston, Texas Southern University, p. 28 (Report No. DOE/CH/10255-T1, prepared for the US Department of Energy, Washington).

WILSON, D.N. (1988b) Cadmium - market trends and influences. In: Cadmium 87. Proceedings of the 6th International Cadmium Conference, London, Cadmium Association, pp. 9-16.

WILSON, R.H., DE EDS, F., & COX, A.J., Jr (1941) Effects of continued cadmium feeding. J. Pharmacol. exp. Ther., **71**: 222-235.

WOLFF, E.W. & PEEL, D.A. (1985) The record of global pollution in polar snow and ice. Nature (Lond.), **313**: 535-540.

WOLNIK, K.A., FRICKE, F.L., CAPER, S.G., BRAUDE, G.L., MEYER, M.W., SATZGER, R.D., & BONNIN, E. (1983) Elements in major raw agricultural crops in the United States. I. Cadmium and lead in lettuce, peanuts, potatoes, soybeans, sweet corn, and wheat. J. agric. food Chem., **31**: 1240-1244.

WOLNIK, K.A., FRICKE, F.L., CAPAR, S.G., MEYER, M.W., & SATZGER, R.D. (1985) Elements in major raw agricultural crops in the United States. 3. Cadmium, lead, and eleven other elements in carrots, field corn, onion, rice, spinach, and tomatoes. J. agric. food Chem., **33**: 807-811.

YAMAGATA, H. & IWASHIMA, K. (1975) Average cadmium intake of the Japanese people. Bull. Inst. Public Health (Tokyo), **24**: 18-24.

YAMAGATA, N. & SHIGEMATSU, I. (1970) Cadmium pollution in perspective. Bull. Inst. Public Health (Tokyo), **19**: 1-27.

YAMAGATA, N., IWASHIMA, K., & NAGAI, T. (1974) [Gastrointestinal absorption of cadmium in normal humans], Tokyo, Japan Public Health Association, pp. 84-85 (Kankyo Hoken Report No. 31) (in Japanese).

YAMAMOTO, Y. (1972) [Present status of cadmium environmental pollution], Tokyo, Japan Public Health Association, pp. 7-12 (Kankyo Hoken Report No. 11) (in Japanese).

YOSHIKAWA, H., KAWAI, K., SUZUKI, Y., NOZAKI, K., & OHSAWA, M. (1975) [Studies on cadmium effects in living organisms: impairment due to inhalation], Tokyo, Japanese Environment Agency, 40 pp (in Japanese).

# RESUME ET CONCLUSIONS

## 1. Identité, propriétés physiques et chimiques et méthodes d'analyse

Il existe plusieurs méthodes pour le dosage du cadmium dans les échantillons biologiques. La plus utilisée est la spectrométrie d'absorption atomique, mais elle nécessite un traitement minutieux de la prise d'essai et une correction pour tenir compte des interférences lorsqu'elle est appliquée à des échantillons de faible teneur en cadmium. Il est tout à fait souhaitable que l'analyse s'accompagne d'un programme d'assurance de la qualité. A l'heure actuelle, il est possible, dans des conditions optimales, de doser environ 0,1 μg/litre dans l'urine et le sang et de 1 à 10 μg/litre dans les aliments et les tissus.

## 2. Sources d'exposition humaine et environnementale

Le cadmium est un élément relativement rare et les méthodes actuelles d'analyse indiquent, dans les divers compartiments de l'environnement, des concentrations beaucoup plus faibles que les mesures antérieures. Pour l'instant, il n'est pas possible de déterminer si l'activité humaine est à l'origine d'un accroissement de la teneur des calottes polaires en cadmium à l'échelle des temps historiques.

La production commerciale de cadmium a commencé au tournant du siècle. La consommation a changé de caractère ces dernières années avec un recul sensible de la galvanoplastie et une utilisation accrue dans la fabrication de batteries et de composants électroniques spéciaux. Dans la plupart des cas, le cadmium est utilisé sous la forme de dérivés peu concentrés, ce qui rend le recyclage indispensable. Les restrictions à l'usage du cadmium imposées par certains pays pourraient avoir des répercussions importantes sur ces applications.

Les activités humaines entraînent la libération de cadmium dans l'air, le sol et l'eau. D'une façon générale, les deux principales sources de contamination sont la production et la consommation de cadmium et d'autres métaux non ferreux ainsi que le rejet de déchets contenant du cadmium. Dans les zones proches de mines ou de fonderies de métaux non ferreux, la contamination par le cadmium est souvent importante.

Plus le sol contient de cadmium, plus la quantité fixée par les plantes est importante. L'exposition humaine par l'intermédiaire des cultures sera donc sensible à toute augmentation de la teneur du sol en cadmium. La fixation par les plantes est plus importante dans les sols de faible pH. Les processus qui acidifient le sol (pluies acides, par exemple) sont donc susceptibles de provoquer une augmentation de la concentration moyenne du cadmium dans les denrées alimentaires. Dans certaines régions du monde, l'utilisation d'engrais phosphatés et les dépôts d'origine atmosphérique constituent une source non négligeable de contamination des terres arables; les boues d'égout peuvent aussi, localement, entraîner une forte pollution. A l'avenir, ces sources risquent d'accroître la contamination du sol et celle des cultures, ce qui débouchera sur une exposition plus importante au cadmium par la voie alimentaire. Dans certaines régions, on peut constater une augmentation de la teneur des aliments en cadmium.

Les animaux et les plantes comestibles qui vivent à l'état sauvage, comme les coquillages, les crustacés et les champignons accumulent naturellement le cadmium. Comme chez l'homme, on constate une augmentation de la concentration en cadmium dans le foie et les reins des chevaux et de certains animaux terrestres vivant à l'état sauvage. La consommation régulière de ces abats peut entraîner une exposition accrue. Dans les reins de certains vertébrés marins on trouve des concentrations assez fortes en cadmium, qui, même si elles sont d'origine naturelle, n'en provoquent pas moins des lésions au niveau de ces organes.

## 3. Concentrations dans l'environnement et exposition humaine

La principale voie d'exposition au cadmium des non fumeurs est la voie alimentaire. Les autres voies sont peu importantes. Chez les fumeurs, l'apport de cadmium par le tabac est notable. Dans les zones contaminées, l'exposition d'origine alimentaire peut atteindre plusieurs centaines de microgrammes par jour. Chez les travailleurs exposés, la principale voie de pénétration est la voie pulmonaire, après inhalation d'air contaminé sur le lieu de travail. Le tabagisme et la consommation d'aliments contaminés ajoutent encore à la charge de cadmium de l'organisme.

## 4. Cinétique et métabolisme chez les animaux de laboratoire et chez l'homme

Les données tirées de l'expérimentation animale et humaine montrent que l'absorption est plus importante au niveau des poumons qu'au niveau des voies digestives. Selon l'espèce chimique en cause, la granulométrie et la solubilité dans les liquides biologiques, le taux d'absorption peut atteindre 50% après inhalation. L'absorption gastro-intestinale dépend du régime alimentaire et de l'état nutritionnel. En particulier, le bilan martial est particulièrement important. En moyenne, le cadmium total contenu dans les aliments est absorbé à hauteur de 5%, avec un intervalle de variation de 1%-20% selon les individus. Il existe un gradient materno-foetal de cadmium. Le cadmium peut également parvenir jusqu'au foetus, mais en faibles quantités malgré son accumulation dans le placenta.

Le cadmium absorbé au niveau des poumons ou des voies digestives s'accumule principalement dans le foie et les reins où il représente plus de la moitié de la charge totale de l'organisme. Plus l'exposition est intense, plus l'accumulation du cadmium dans le foie est importante. En principe, l'excrétion est lente et la période biologique du cadmium dans les muscles, les reins, le foie et l'organisme dans son ensemble, est très longue, de l'ordre de plusieurs décennies. La teneur en cadmium de la plupart des tissus augmente avec l'âge. C'est en général dans le cortex rénal que la concentration est la plus élevée, mais en cas d'exposition excessive, elle peut l'être encore plus dans le foie. Chez les personnes exposées atteintes de lésions rénales, il y a augmentation de l'excrétion urinaire du cadmium de sorte que la période biologique pour l'ensemble de l'organisme est raccourcie. Du fait des lésions, le rein perd son cadmium et ces malades finissent par présenter une concentration rénale de cadmium plus faible que les individus en bonne santé soumis à la même exposition.

La métallothionéine est une protéine qui joue un rôle important dans le transport et le stockage du cadmium et d'autres métaux. Le cadmium est capable d'induire la synthèse de cette protéine dans de nombreux organes, notamment le foie et le rein. En fixant le cadmium intracellulaire, la métallothionéine protège les tissus contre les effets toxiques de ce métal. Il est possible, par conséquent, que le cadmium non lié à la métallothionéine ait une responsabilité dans les lésions tissulaires. On ignore quels peuvent être les autres complexes du cadmium présents dans les liquides biologiques.

L'excrétion urinaire du cadmium dépend de divers facteurs: charge totale de l'organisme, exposition récente et lésions rénales. Chez les individus peu exposés, la concentration urinaire du cadmium dépend principalement de la charge de l'organisme. En cas de lésions rénales dues au cadmium, ou même sans lésions de ce genre mais en présence d'une exposition excessive, il y a augmentation de l'excrétion urinaire. Les individus exposés au cadmium en excrètent davantage dans leurs urines lorsqu'ils présentent une protéinurie. Après cessation d'une exposition intense, le taux urinaire décroît, même s'il y a persistance des lésions rénales. Il faut donc prendre plusieurs facteurs en considération pour interpréter le cadmium urinaire. L'excrétion par la voie digestive est sensiblement équivalente à l'excrétion urinaire mais elle est difficile à mesurer. L'excrétion par d'autres voies (lactation, sueur ou passage transplacentaire) est négligeable.

La teneur des matières fécales en cadmium est un bon indicateur d'une ingestion récente en l'absence d'exposition par la voie respiratoire. Le cadmium sanguin est présent principalement dans les hématies, la concentration plasmatique étant très faible. Il existe au moins deux compartiments dans le sang, l'un qui correspond à une exposition récente, avec une demi-vie de 2-3 mois et l'autre, qui est probablement lié à la charge totale de l'organisme et se caractérise par une demi-vie de plusieurs années.

## 5. Effets sur les animaux de laboratoire

Une forte exposition par la voie respiratoire entraîne un oedème mortel du poumon. Après injection d'une seule dose, apparaissent des lésions testiculaires, une nécrose ovarienne, des lésions hépatiques et une atteinte des petits vaisseaux. L'ingestion de fortes doses provoque des lésions de la muqueuse gastrique et intestinale.

Une exposition prolongée par la voie respiratoire ou une administration intratrachéenne entraîne des altérations pulmonaires de nature inflammatoire ainsi qu'une fibrose et donne au tissu pulmonaire un aspect qui évoque l'emphysème. L'administration prolongée par voie orale ou parentérale affecte principalement le rein mais elle a aussi des effets sur le foie et les systèmes hématopoiétique, immunitaire et cardio-vasculaire ainsi que sur le squelette. Chez certaines espèces et dans des conditions déterminées, on a provoqué une hypertension et constaté des effets sur le squelette. C'est le stade de la gestation où se produit l'exposition qui conditionne les effets tératogènes et les lésions placentaires et il peut y avoir interaction avec le zinc.

Ce sont les effets aigus produits par l'inhalation du cadmium ainsi que sa néphrotoxicité chronique qui sont les plus importants du point de vue de l'exposition humaine. En cas d'exposition prolongée, c'est le rein qui est l'organe critique. Les effets sont caractérisés par une lésion des cellules tubulaires entraînant une insuffisance tubulaire parfois accompagnée d'insuffisance glomérulaire. L'insuffisance tubulaire a pour conséquence une perturbation du métabolisme du calcium et de la vitamine D. Selon certains travaux, ces troubles pourraient provoquer une ostéomalacie ou une ostéoporose. Cependant ces résultats n'ont pas été confirmés par d'autres études. On ne peut exclure un effet direct du cadmium sur la minéralisation de l'os. Chez l'animal de laboratoire, les effets toxiques du cadmium dépendent de certains facteurs génétiques et nutritionnels, des interactions avec d'autres métaux, en particulier le zinc, et d'un premier traitement éventuel par le cadmium susceptible d'avoir stimulé la synthèse de métallothionéine.

En 1976 et 1987, le Centre international de recherche sur le cancer a admis posséder suffisamment de preuves que l'injection de chlorure, de sulfate, de sulfure et d'oxyde de cadmium pouvait entraîner l'apparition d'un sarcome local chez le rat et, dans le cas des deux premiers composés, de tumeurs testiculaires interstitielles chez ce même animal et chez la souris. Toutefois, il a considéré que les études basées sur l'administration par voie orale ne permettaient pas de procéder à une évaluation. Lors d'études au cours desquelles on a fait respirer à des rats des aérosols de sulfate de cadmium, des vapeurs d'oxyde de cadmium et de la poussière de sulfate de cadmium, on a observé une forte incidence de cancers primitifs du poumon, avec probablement une relation entre la dose et la réponse. Toutefois, ces résultats n'ont pu être reproduits chez d'autres espèces. Les travaux relatifs aux effets génotoxiques du cadmium ont donné des résultats contradictoires.

## 6. Effets sur l'homme

Une forte exposition à des vapeurs d'oxyde de cadmium par la voie respiratoire entraîne une pneumopathie aiguë, accompagnée d'un oedème du poumon qui peut être mortel. L'ingestion de grandes quantités de sels solubles de cadmium provoque une gastro-entérite aiguë.

A la suite d'une exposition professionnelle prolongée au cadmium, on a observé de graves effets chroniques, principalement au niveau des poumons et des reins. On a également observé une néphrotoxicité chronique dans la population générale.

Les altérations pulmonaires consécutives à une exposition professionnelle intense sont essentiellement caractérisées par une obstruction des voies aériennes. Si l'exposition se poursuit, les légers troubles ventilatoires initiaux peuvent déboucher sur une insuffisance respiratoire. On a observé un accroissement de la mortalité par pneumopathie obstructive chez des travailleurs fortement exposés, comme cela se produisait auparavant.

L'accumulation de cadmium dans le cortex rénal entraîne des troubles de la fonction tubulaire et une réabsorption insuffisante, par exemple, des protéines, du glucose et des acides aminés. L'accroissement de l'excrétion urinaire des protéines de faible masse moléculaire est un signe caractéristique de l'insuffisance tubulaire. Parfois il y a aussi baisse du taux de filtration glomérulaire. L'augmentation du cadmium urinaire est corrélée avec une protéinurie de faible masse moléculaire et, en l'absence d'exposition, peut servir d'indicateur de l'atteinte rénale. Dans les cas graves, les effets tubulaires et glomérulaires s'ajoutent s'accompagnant parfois d'une élévation du taux sanguin de créatinine. Chez la plupart des travailleurs et autres personnes, la protéinurie due à une néphropathie cadmique est irréversible.

Parmi les autres effets, on peut citer les troubles du métabolisme calcique, l'hypercalciurie et la formation de calculs rénaux. Une exposition intense au cadmium peut, selon toute probabilité lorsqu'elle s'accompagne d'autres facteurs comme une carence nutritionnelle, provoquer l'apparition d'une ostéoporose et/ou d'une ostéomalacie.

On est fondé à penser qu'une exposition professionnelle prolongée au cadmium peut favoriser l'apparition d'un cancer du poumon, mais la présence de facteurs de confusion ne facilite pas l'interprétation des observations effectuées sur les travailleurs exposés. En ce qui concerne le cancer de la prostate, les données ne sont pas concluantes et ne confirment pas, en tout cas, l'hypothèse antérieure d'une relation de cause à effet.

A l'heure actuelle, on ne possède pas de preuve convaincante que le cadmium provoque une hypertension essentielle. La plupart des données contredisent cette hypothèse et rien n'indique un accroissement de la mortalité par maladie cardio-vasculaire ou accident vasculaire cérébral chez les personnes exposées.

D'après les résultats d'études relatives à des groupes exposés de par leur profession ou simplement du fait de leur environnement général, il semble que la prévalence des effets néphrotoxiques soit liée à la durée et à l'intensité de l'exposition.

Chez des ouvriers de l'industrie du cadmium, on a signalé, après 10 à 20 ans d'exposition à des concentrations de l'ordre de 20-50 μg par mètre cube, une augmentation de la prévalence des cas de protéinurie à faible masse moléculaire.

Dans des zones polluées, où l'on évalue l'apport de cadmium par voie orale à environ 140-260 μg/jour, on a observé des effets du genre protéinurie à faible masse moléculaire chez des sujets exposés pendant une longue période. On trouvera à la section 8 une estimation plus précise de la relation dose-réponse.

## 7. Evaluation des risques pour la santé humaine

### 7.1 Conclusions

On estime que le rein est l'organe cible tant dans la population générale que chez les groupes professionnellement exposés. Une exposition prolongée par inhalation entraîne l'apparition d'un syndrome respiratoire obstructif chez certains groupes professionnels. Certains détails incitent à penser que cette exposition au cadmium pourrait favoriser l'apparition d'un cancer du poumon, mais les observations effectuées sur des travailleurs exposés sont difficiles à interpréter en raison de la présence de facteurs de confusion.

#### 7.1.1 Population générale

Pour la plupart des individus, les aliments constituent la principale voie d'exposition au cadmium. Dans la plupart des régions non polluées par ce métal, l'apport alimentaire journalier est de l'ordre de 10-40 μg. Dans les zones polluées, il peut atteindre plusieurs centaines de microgrammes par jour. Dans les zones non polluées, l'apport dû au tabac peut être égal à l'apport alimentaire chez les gros fumeurs.

D'après un modele biologique, on estime qu'il existe une association entre l'exposition au cadmium et l'excrétion urinaire de protéines de faible masse moléculaire chez les sujets qui absorbent pendant toute leur vie une dose journalière d'environ 140-260 μg de cadmium, ce qui correspond à une dose cumulée d'environ 2000 mg ou davantage.

### 7.1.2 *Groupes professionnellement exposés*

Dans ce cas, l'exposition est essentiellement respiratoire, mais il s'y ajoute l'apport alimentaire et tabagique. La teneur totale de l'air en cadmium varie selon les pratiques en matière d'hygiène industrielle et selon le lieu de travail. Il existe une relation de type exposition-réponse entre la teneur de l'air en cadmium et la protéinurie de faible masse moléculaire. La prévalence de cette protéinurie peut augmenter après 10 à 20 ans d'exposition à des concentrations de cadmium de l'ordre de 20-50 μg par mètre cube. *In vivo*, le dosage du cadmium dans les reins et le foie de sujets plus ou moins exposés a montré que 10 % environ des travailleurs dont le cortex rénal contenait 200 mg/kg de cadmium et 50 % de ceux chez qui cette concentration atteignait 300 mg/kg, feraient un jour ou l'autre une protéinurie par insuffisance tubulaire.

# RESUMEN Y CONCLUSIONES

## 1. Identidad, propiedades físicas y químicas, y métodos analíticos

Se dispone de varios métodos para determinar el cadmio presente en el material biológico. El más utilizado es la espectrometría de absorción atómica, aunque el análisis de muestras con concentraciones bajas de cadmio exige un tratamiento cuidadoso de las muestras y correcciones para tener en cuenta la interferencia. Se recomienda encarecidamente acompañar el análisis con un programa de garantía de la calidad. Actualmente, en circunstancias ideales pueden determinarse concentraciones de alrededor de 0,1 μg/litro en la orina y la sangre y de 1-10 μg/kg en alimentos y muestras de tejidos.

## 2. Fuentes de exposición humana y ambiental

El cadmio es un elemento relativamente raro; los procedimientos analíticos actuales indican que las concentraciones del metal en el medio ambiente son mucho más bajas que las obtenidas en medidas anteriores. Hoy en día no es posible determinar si la actividad humana ha provocado un aumento histórico de los niveles de cadmio en los casquetes polares.

La producción comercial de cadmio comenzó a principios de este siglo. La pauta de consumo de cadmio se ha modificado en los últimos años debido al notable descenso del uso de la galvanoplastia y al importante aumento de la producción de baterías y de las aplicaciones electrónicas especializadas. En las principales aplicaciones del cadmio éste se utiliza en forma de compuestos que se hallan presentes en bajas concentraciones; ello obstaculiza el reciclaje del metal. Las restricciones impuestas por algunos países a ciertas aplicaciones del cadmio pueden tener un efecto generalizado en esas aplicaciones.

El cadmio se libera al aire, los suelos y las aguas debido a la actividad humana. En general, las dos fuentes principales de contaminación son la producción y el consumo de cadmio y de otros metales no ferrosos y la evacuación de desechos que contienen cadmio. Las zonas próximas a minas no ferrosas y fundiciones suelen estar muy contaminadas por cadmio.

Al aumentar el contenido de cadmio del suelo, aumenta la absorción del metal por las plantas; la exposición humana a partir de las cosechas agrícolas está por tanto sometida a los aumentos del contenido de cadmio del suelo. Dado que la absorción por las plantas desde el suelo es mayor cuando el pH de éste es bajo, los procesos que acidifican el suelo (por ejemplo, las lluvias ácidas) pueden aumentar las concentraciones medias de cadmio en los alimentos. La aplicación de fertilizantes a base de fosfato y la deposición atmosférica son fuentes importantes de aportación de cadmio a las tierras cultivables en ciertas zonas del mundo; los fangos de alcantarillado también pueden ser una fuente de importancia a nivel local. Estas fuentes pueden, en el futuro, aumentar los niveles de cadmio en el suelo y con ello en las cosechas, lo que a su vez puede acrecentar la exposición al cadmio en la dieta. En ciertas zonas, se ha demostrado que está aumentando el contenido de cadmio en los alimentos.

Ciertos organismos comestibles de vida libre como los mariscos, los crustáceos y los hongos son acumuladores naturales de cadmio. Como en el caso del ser humano, se observan niveles mayores de cadmio en el hígado y el riñón de los caballos y de algunos animales terrestres silvestres. El consumo habitual de estos alimentos puede aumentar la exposición. Ciertos vertebrados marinos contienen concentraciones notablemente elevadas de cadmio en el riñón, fenómeno que, aunque se considera de origen natural, se ha vinculado a signos de lesiones renales en esos organismos.

## 3. Niveles ambientales y exposición humana

La principal fuente de exposición al cadmio en la población general no fumadora son los alimentos; la proporción de cadmio que se absorbe por otras vías es pequeña. El tabaco es una importante fuente de absorción de cadmio en los fumadores. En las zonas contaminadas, la exposición al cadmio por los alimentos puede alcanzar varios cientos de $\mu$g/día. En los trabajadores expuestos, la absorción pulmonar de cadmio por inhalación en el lugar de trabajo es la principal vía de exposición. También puede aumentar la absorción por la contaminación de los alimentos y por el consumo de tabaco.

## 4. Cinética y metabolismo en animales de experimentación y en el ser humano

Los datos obtenidos en animales de experimentación y en el ser humano han demostrado que la absorción pulmonar es mayor que la gastrointestinal. Atendiendo a la especiación química, el tamaño de las partículas y la solubilidad en fluidos biológicos, puede absorberse hasta el 50% del compuesto de cadmio inhalado. La absorción gastrointestinal de cadmio depende del tipo de dieta y del estado nutricional. El estado nutricional respecto del hierro parece revestir particular importancia. Aunque en promedio se absorbe el 5% de la ingesta oral total de cadmio, los valores individuales varían entre menos del 1% hasta más del 20%. Existe un gradiente maternofetal de cadmio. Aunque se acumula en la placenta, la transferencia al feto es baja.

El cadmio absorbido en los pulmones o el tracto gastrointestinal se almacena principalmente en el hígado y el riñón, donde se deposita más de la mitad de la carga corporal. Al aumentar la intensidad de la exposición, aumenta la proporción del cadmio absorbido que se almacena en el hígado. En el ser humano la excreción suele ser lenta y la semivida biológica es muy larga (decenios) en el músculo, el riñón, el hígado y el organismo entero. Las concentraciones de cadmio en la mayoría de los tejidos aumentan con la edad. Aunque las concentraciones más elevadas suelen encontrarse en la corteza renal, con exposiciones excesivas pueden producirse concentraciones mayores en el hígado. En las personas expuestas que padecen lesiones renales, aumenta la excreción urinaria de cadmio con lo que se reduce la semivida en el organismo entero. Las lesiones renales producen pérdidas del cadmio contenido en el riñón, y las concentraciones renales acaban con el tiempo siendo inferiores a las observadas en personas con un grado de exposición similar pero sin lesiones renales.

La metalotioneína es una importante proteína de transporte y almacenamiento de cadmio y otros metales. El cadmio puede inducir la síntesis de metalotioneína en muchos órganos, en particular el hígado y el riñón. La unión del cadmio intracelular a la metalotioneína en los tejidos protege contra la toxicidad del metal. El cadmio libre puede por tanto tener una función en la patogenia de las lesiones tisulares debidas a ese metal. Se desconoce la especiación de otros complejos de cadmio en los tejidos o en los fluidos biológicos.

La excreción urinaria de cadmio guarda relación con la carga corporal, la exposición reciente y la lesión renal. En personas poco expuestas, el nivel de cadmio en la orina depende principalmente de la carga corporal. Una vez que se ha producido la lesión renal inducida por el cadmio, o incluso en ausencia de lesión renal si la exposición es excesiva, aumenta la excreción urinaria. En las personas expuestas al cadmio que padecen proteinuria la excreción de cadmio suele ser mayor que en las que no padecen proteinuria. Cuando cesa la exposición intensa, el nivel de cadmio en la orina desciende aunque persista la lesión renal. En la interpretación de la presencia de cadmio en la orina hay que tener en cuenta, pues, varios factores. La excreción gastrointestinal es aproximadamente igual a la urinaria pero no puede medirse fácilmente. Otras vías excretoras como la leche, el sudor o la transferencia placentaria son insignificantes.

El nivel de cadmio en las heces es un buen indicador de la ingesta diaria reciente a partir de los alimentos en ausencia de exposición por inhalación. En la sangre, el cadmio aparece principalmente en los glóbulos rojos y las concentraciones en el plasma son muy bajas. Existen al menos dos compartimentos en la sangre, uno referido a la exposición reciente, con una semivida de alrededor de 2-3 meses, y otro probablemente relacionado con la carga corporal, con una semivida de varios años.

## 5. Efectos en mamíferos de laboratorio

Las exposiciones elevadas por inhalación provocan edema pulmonar letal. La inyección de una sola dosis elevada produce necrosis en el testículo y en el ovario no ovulante, lesiones hepáticas y lesiones en los vasos de menor tamaño. La administración oral de dosis elevadas produce lesiones en la mucosa gástrica e intestinal.

La exposición por inhalación prolongada y la administración intratraqueal producen modificaciones crónicas de tipo inflamatorio en los pulmones, fibrosis y fenómenos indicativos de enfisema. La administración parenteral u oral prolongada afecta principalmente al riñón aunque también al hígado y a los sistemas hematopoyético, inmunitario, esquelético y cardiovascular. En ciertas especies y en determinadas condiciones se han inducido efectos esqueléticos e hipertensión. La aparición de efectos teratogénicos y lesiones placentarias depende de la fase gestacional en que se produzca la exposición y puede entrañar interacción con el zinc.

En cuanto a la exposición humana, lo más notable son los efectos agudos por inhalación en el pulmón y los efectos renales crónicos. Tras la exposición prolongada, el riñón es el órgano crítico. Los efectos en este órgano se caracterizan por disfunción tubular y lesiones en las células tubulares, si bien pueden producirse también disfunciones glomerulares. Una de las consecuencias de la disfunción tubular renal es la alteración del metabolismo del calcio y de la vitamina D. Según algunos estudios, ello ha producido casos de osteomalacia y/o osteoporosis, pero esos efectos no se han confirmado en otros estudios. No debe excluirse el efecto directo del cadmio en la mineralización ósea.

Los efectos tóxicos del cadmio en animales de experimentación están sometidos a la influencia de factores genéticos y nutricionales, las interacciones con otros metales, en particular el zinc, y el pretratamiento con cadmio, que puede guardar relación con la inducción de la metalotioneína.

En 1976 y 1987, el Centro Internacional de Investigaciones sobre el Cáncer consideró suficientes las pruebas de que el cloruro, el sulfato, el sulfuro y el óxido de cadmio pueden producir sarcomas en el lugar de inyección en la rata y, en el caso de los dos primeros compuestos, inducir tumores en las células intersticiales del testículo en la rata y el ratón, pero consideró que los estudios de administración oral eran insuficientes para la evaluación. En estudios de inhalación prolongada en ratas expuestas a aerosoles de sulfato de cadmio, vapores de óxido de cadmio y polvos de sulfato de cadmio se observó una elevada incidencia de cáncer primario del pulmón con pruebas de proporcionalidad entre la dosis y la respuesta. Hasta el momento, sin embargo, esa observación no se ha confirmado en otras especies. Los estudios de los efectos genotóxicos del cadmio han dado resultados discordantes.

## 6. Efectos en el ser humano

La exposición intensa por inhalación de vapores de óxido de cadmio produce neumonitis aguda con edema pulmonar, que puede ser letal. La ingestión de dosis elevadas de sales solubles de cadmio produce gastroenteritis aguda.

La exposición ocupacional prolongada al cadmio ha producido efectos crónicos graves, principalmente en el pulmón y el riñón. También se han observado efectos renales crónicos en la población general.

Los cambios pulmonares observados tras una intensa exposición ocupacional se caracterizan principalmente por la aparición de afecciones crónicas obstructivas de las vías aéreas. Los primeros cambios leves en las pruebas de la función ventilatoria pueden avanzar, si prosigue la exposición al cadmio, hasta insuficiencia respiratoria. Se ha observado una mayor tasa de mortalidad por enfermedad pulmonar obstructiva en trabajadores sometidos a exposiciones intensas, al igual que en otras épocas.

La acumulación de cadmio en la corteza renal produce disfunción tubulorrenal con trastornos de la reabsorción de proteínas, glucosa y aminoácidos, entre otros. Un signo característico de la disfunción tubular es la mayor excreción de proteínas de bajo peso molecular en la orina. En algunos casos, disminuye la tasa de filtración glomerular. El aumento de la concentración de cadmio en la orina está correlacionado con la presencia de proteínas de bajo peso molecular en la orina y, en ausencia de exposición aguda al cadmio, puede servir como indicador de efectos renales. En los casos más graves se combinan los efectos tubulares y glomerulares, con aumento del nivel de creatinina en la sangre en algunos casos. Para la mayoría de los trabajadores y de las personas expuestas al medio ambiente general, la proteinuria inducida por el cadmio es irreversible.

Entre otros efectos figuran los trastornos del metabolismo del calcio, la hipercalciuria y la formación de cálculos renales. La exposición intensa al cadmio, con toda probabilidad combinado con otros factores como carencias nutricionales, puede llevar a la aparición de osteoporosis y/o osteomalacia.

Hay pruebas de que la exposición profesional prolongada al cadmio puede contribuir a la aparición de cáncer del pulmón aunque las observaciones obtenidas en trabajadores expuestos han sido difíciles de interpretar a causa de factores que inducen a confusión. En el caso del cáncer de la próstata, las pruebas obtenidas hasta la fecha no son concluyentes pero no apoyan la existencia de una relación causal, indicada en estudios anteriores.

Actualmente no hay pruebas convincentes de que el cadmio sea agente etiológico de la hipertensión esencial. La mayor parte de los datos indican que no se debe al cadmio el aumento de la tensión y no hay pruebas de que la mortalidad por enfermedades cardiovasculares o cerebrovasculares sea mayor.

Los datos obtenidos en estudios de grupos de trabajadores expuestos y de grupos expuestos en el medio ambiente general demuestran que existe una relación entre los niveles de exposición, la duración de ésta y la prevalencia de los efectos renales.

Se ha comunicado una mayor prevalencia de la proteinuria de bajo peso molecular en trabajadores del cadmio tras 10-20 años de exposición a niveles del metal de aproximadamente 20-50 $\mu g/m^3$.

En zonas contaminadas del medio general, en las que la ingesta de cadmio estimada ha sido de 140-260 μg/día, se han observado efectos en forma de aumento de la cantidad de proteínas de bajo peso molecular en la orina en algunos individuos tras una exposición prolongada. En la sección 8 se dan estimaciones más precisas de la relación dosis-respuesta.

## 7. Evaluación de los riesgos para la salud humana

### 7.1 Conclusiones

Se considera que el riñón es el órgano diana crítico en la población general así como en las poblaciones expuestas profesionalmente. Las enfermedades crónicas obstructivas de las vías respiratorias están asociadas a la exposición profesional prolongada e intensa por inhalación. Hay pruebas de que esa exposición al cadmio puede contribuir al desarrollo de cáncer del pulmón aunque las observaciones en trabajadores expuestos han sido difíciles de interpretar a causa de la presencia de factores que inducen a confusión.

#### 7.1.1 Población general

El cadmio presente en los alimentos es la principal fuente de exposición para la mayoría de las personas. En la mayoría de las zonas no contaminadas con cadmio las ingestas diarias medias con los alimentos se encuentran entre 10 y 40 μg. En zonas contaminadas se ha observado que alcanza varios cientos de μg al día. En zonas no contaminadas, la absorción debida al consumo de tabaco puede igualar la ingestión de cadmio a partir de los alimentos.

Basandose en un modelo biologico, se ha estimado que con una ingesta diaria de 140-260 μg de cadmio durante toda la vida, o una ingesta acumulativa de unos 2000 mg o más, se produce en el ser humano una asociación entre la exposición al cadmio y una mayor excreción de proteínas de bajo peso molecular en la orina.

#### 7.1.2 Población expuesta profesionalmente

La exposición ocupacional al cadmio se produce principalmente por inhalación aunque comprende ingestas suplementarias con los alimentos y el tabaco. El nivel total de cadmio en el aire varía según las prácticas de higiene industrial y el tipo de lugar de trabajo. Existe una relación exposición-respuesta entre los niveles de cadmio en el aire y la proteinuria. Puede aumentar la prevalencia de la proteinuria de bajo peso molecular en trabajadores a los 10-20 años de exposición a niveles de cadmio de unos 20-50 $\mu g/m^3$. La medida *in vivo* del cadmio en el riñón y el hígado de personas con distintos niveles de exposición al metal ha demostrado que alrededor del 10% de los trabajadores con un nivel en la corteza renal de 200 mg/kg y aproximadamente el 50% de las personas con un nivel en la corteza renal de 300 mg/kg tendrían proteinuria tubulorrenal.